Tae Eun Kim · Marcelo Milrad ·
Inmaculada Remolar

Editors

Integrating Emerging Technologies into Education and Training: Proceedings of the 2nd ETELT 2024 Workshop

Methodologies and Intelligent Systems
for Technology Enhanced Learning,
14th International Conference

 Springer

Editors
Tae Eun Kim
University of Tromsø (UiT) – The Arctic
University of Norway
Tromsø, Norway

Marcelo Milrad
Department of Computer Science and Media
Technology
Linnaeus University
Växjö, Sweden

Inmaculada Remolar
Institute of New Imaging Technologies
Jaume I University
Castelló de la Plana, Spain

ISSN 2367-3370   ISSN 2367-3389 (electronic)
Lecture Notes in Networks and Systems
ISBN 978-3-031-84169-9   ISBN 978-3-031-84170-5 (eBook)
https://doi.org/10.1007/978-3-031-84170-5

This Springer imprint is published by the registered company Springer Nature Switzerland AG
The registered company address is: Gewerbestrasse 11, 6330 Cham, Switzerland

If disposing of this product, please recycle the paper.

Lecture Notes in Networks and Systems 1274

The series "Lecture Notes in Networks and Systems" publishes the latest developments in Networks and Systems—quickly, informally and with high quality. Original research reported in proceedings and post-proceedings represents the core of LNNS.

Volumes published in LNNS embrace all aspects and subfields of, as well as new challenges in, Networks and Systems.

The series contains proceedings and edited volumes in systems and networks, spanning the areas of Cyber-Physical Systems, Autonomous Systems, Sensor Networks, Control Systems, Energy Systems, Automotive Systems, Biological Systems, Vehicular Networking and Connected Vehicles, Aerospace Systems, Automation, Manufacturing, Smart Grids, Nonlinear Systems, Power Systems, Robotics, Social Systems, Economic Systems and other. Of particular value to both the contributors and the readership are the short publication timeframe and the worldwide distribution and exposure which enable both a wide and rapid dissemination of research output.

The series covers the theory, applications, and perspectives on the state of the art and future developments relevant to systems and networks, decision making, control, complex processes and related areas, as embedded in the fields of interdisciplinary and applied sciences, engineering, computer science, physics, economics, social, and life sciences, as well as the paradigms and methodologies behind them.

Indexed by SCOPUS, EI Compendex, INSPEC, WTI Frankfurt eG, zbMATH, SCImago.

All books published in the series are submitted for consideration in Web of Science.

For proposals from Asia please contact Aninda Bose (aninda.bose@springer.com).

Preface

Emerging technologies encompassing artificial intelligence, digitalization, robotics and the Internet of Things (IoT) are driving transformative changes in education. By leveraging innovations such as Virtual Reality (VR), Augmented Reality (AR) and AI-powered virtual assistants, these technologies foster immersive and experiential learning environments that could significantly enhance engagement and adaptability in the learning process, particularly in disciplines such as healthcare and engineering, where students can simulate complex tasks to gain both theoretical knowledge and practical skills.

AI-powered adaptive learning enables personalized and self-paced education for students and also supports teachers in gaining insights into learning patterns, strengths and areas needing improvement, which allows for more informed educational guidance. Additionally, mobile devices and online platforms have expanded access to allow flexible and continuous learning. However, Technology Enhanced Learning (TEL) presents both promising opportunities and challenges, including technology dependence, digital divides, cyber security risks and skill gaps, all of which must be responsibly considered to fully harness TEL's potential.

This workshop is jointly organized by three EU Horizon projects that have been funded by the Horizon Europe program and under the topic "Integration of emerging new technologies into education and training" and focuses on current research trends, views and results related to emerging technology-enhanced education and training in various domains. It presented and elaborated the state-of-the-art research on related methodologies and intelligent systems and discussed their potential applications in different learning and industrial contexts.

All papers underwent a peer-review selection: each paper was assessed by two/three different reviewers, from an international panel. All 11 quality papers (from a total of 20 papers), with authors coming from various countries, have been presented and included in the present volume.

This workshop is organized under the frame of the 14th International Conference on Methodologies and Intelligent Systems for Technology Enhanced Learning (MIS4TEL) which is organized by the BISITE Research Group of the University of Salamanca (Spain). We would like to thank all the contributing authors, the members of the Program Committee, the reviewers, the sponsors and the Organizing Committee for their highly valuable work. Thanks for your help—ETELT & MIS4TEL 2024 would not exist without your contribution.

Marcelo Milrad
Inmaculada Remolar
Tae Eun Kim

Organization of ETELT 2024

Organizing Committee

Marcelo Milrad   Linnaeus University, Sweden
Inmaculada Remolar   Universitat Jaume I, Spain
Tae-Eun Kim   UiT The Arctic University of Norway, Norway

Program Committee

Federica Caruso	University of L'Aquila, Italy
Tania Di Mascio	DISIM, University of L'Aquila, Italy
Rubén García Vidal	Universitat Jaume I, Spain
Garofalakis George	Uni Systems, Greece
Gerardo Herrera Gutiérrez	Universidad de Valencia, Spain
Christothea Herodotou	Institute of Educational Technology, UK
Nikos Karacapilidis	University of Patras, Greece
Tae-Eun Kim	University of Tromsø, Norway
Marcelo Milrad	Linnaeus University, Sweden
Sofia Papavlasopoulou	Norwegian University of Science and Technology, Norway
Kai Pata	Tallinn University, Estonia
Lokukaluge Prasad Perera	UiT The Arctic University of Norway, Norway
Inmaculada Remolar	Universidad Jaume I, Spain
Margarida Romero	Université de Nice Sophia Antipolis, France
Carlos Santos	University of Aveiro, Portugal
Hans Joachim Schramm	WU Vienna University, Austria
Sagun Shrestha	The Open University, OK
László Szécsi	Budapest University of Technology and Economics, Hungary
Marco Temperini	Sapienza-Università di Roma, Italy
Dimitris Tsakalidis	Novelcore, Greece
Pierpaolo Vittorini	University of L'Aquila, Italy
Feiran Zhang	Norwegian University of Science and Technology, Norway

Organization of MIS4TEL 2024

http://www.mis4tel-conference.net/

General Chair

Christothea Herodotou   The Open University, UK

Technical Program Chairs

Sofia Papavlasopoulou   Norwegian University of Science and Technology,
                                     Norway
Carlos Santos   University of Aveiro, Portugal

Workshop Chairs

Zuzana Kubincová   Comenius University in Bratislava, Slovakia
Dalila Durães   University of Minho, Portugal

Scientific Chairs

Marcelo Milrad   Linnaeus University, Sweden
Nuno Otero   University of Greenwich, UK

Publicity Chair

María Cruz Sánchez-Gómez   University of Salamanca, Spain

Steering Committee Representatives

Pierpaolo Vittorini   University of L'Aquila, Italy
Rosella Gennari   Free University of Bozen-Bolzano, Italy
Tania Di Mascio   University of L'Aquila, Italy
Marco Temperini   Sapienza University of Rome, Italy
Fernando De la Prieta   University of Salamanca, Spain

Organizing Committee

Fernando De la Prieta (Chair)	University of Salamanca, Spain
Sara Rodríguez González	University of Salamanca, Spain
Javier Prieto Tejedor	University of Salamanca, Spain
Ricardo S. Alonso Rincón	AIR Institute, Spain
Alfonso González Briones	University of Salamanca, Spain
Pablo Chamoso Santos	University of Salamanca, Spain
Javier Parra	University of Salamanca, Spain
Belén Pérez Lancho	University of Salamanca, Spain
Ana Belén Gil González	University of Salamanca, Spain
Ana De Luis Reboredo	University of Salamanca, Spain
Angélica González Arrieta	University of Salamanca, Spain
Angel Luis Sánchez Lázaro	University of Salamanca, Spain
Emilio S. Corchado Rodríguez	University of Salamanca, Spain
Guillermo Hernández	University of Salamanca, Spain
Liliana Durón	University of Salamanca, Spain
Laura Grande	University of Salamanca, Spain
Carmen Losada	AIR Institute, Spain
Albano Carrera González	AIR Institute, Spain
Marta Plaza Hernández	University of Salamanca, Spain
Beatriz Bellido	University of Salamanca, Spain
María Alonso	University of Salamanca, Spain
Diego Valdeolmillos	AIR Institute, Spain
Carlos Álvarez	AIR Institute, Spain

Contents

Gaps in Tertiary Education Institutions to Facilitate Practice-Based E-Learning with Disruptive Learning Technologies

Mairi Matrov[1]([✉]), Kai Pata[1], Terje Väljataga[1], Katrin Karu[1],
Águeda Gómez-Cambronero[2], Rubén García-Vidal[2], Inmaculada Remolar[2],
Katerina Theodoridou[3], László Szécsi[4], Elena Llorca-Asensi[5], and Tania Trayanova[6]

[1] Tallinn University, Narva mnt 25, 10712 Tallinn, Estonia
`mairi@tlu.ee`
[2] Jaume I University (UJI), Avinguda de Vicent Sos Baynat, s/n, 12006 Castelló de la Plana, Castelló, Spain
[3] Center for Social Innovation (CSI), 62, Rigainis Street, 1st Floor, 1010 Nicosia, Cyprus
[4] Budapesti Muszaki es Gazdasagtudomanyi Egyetem (BME), MŰEGYETEM RAKPART 3, BUDAPEST 1111, Hungary
[5] University of Alicante, Carr. de San Vicente del Raspeig, s/n, 03690 San Vicente del Raspeig, Alicante, Spain
[6] Innogrowth European Association for Innovation and Growth (INN), 14 Tsar Osvoboditel blvd, Sofia 1000, Bulgaria

Abstract. This paper describes the maturity of tertiary education institutions (HEIs) in integrating practice-based e-learning with disruptive technologies (DT). It identifies gaps that hinder the transition to DT facilitated practice-based elearning. Within the scope of the project e-DIPLOMA, a self-evaluation questionnaire was developed and an online survey was carried out in the project partner countries. Collected dataset incorporated responses from 87 technical and didactic support personnel, 327 educators and 433 students from 92 tertiary education institutions. To describe the current state of maturity of HEIs to facilitate practical elearning with DT, mean values of the key components were calculated. A t-test was used to identify statistically significant differences in responses between groups of respondents. The use of DT in practice based elearning is hindered by the lack of sufficient infrastructure in HEIs, with further constraints arising from limited lecturer involvement in training, communities of practice, and in external collaborations. The findings reveal that educators have limited experience with teaching using DT. A rather positive attitude from all participant groups towards the positive impact of engaging DT both learning and sustainability perspectives..

Keywords: Digital Maturity · Disruptive Learning Technologies · Practice-based E-learning

T. E. Kim et al. (Eds.): MIS4TEL 2024, LNNS 1274, pp. 1–10, 2025.
https://doi.org/10.1007/978-3-031-84170-5_1

1 Introduction

The campus closures caused by the Covid-19 pandemic, revealed the gap of the educational institutions' digital maturity to conduct the practice based e-learning, and practical competences acquisition was disrupted. The increased reliance on digital technologies during the lockdown accelerated the tertiary education institutions' (HEIs) adoption of disruptive learning technologies (DT), such as artificial intelligence (AI), augmented reality (AR), virtual reality (VR) and learning robots. The urgency to integrate new DT into pedagogical settings is further amplified by the fast-evolving job market, necessitating an alignment of educational content with industry needs [1, 2]. Echoing the need for digital adaptation, government bodies and public sectors are calling for fostering the development of a high-performing digital education ecosystem and enhancing digital skills and competences for the digital transformation [3]. Moreover, studies of pandemic experiences also revealed Generation Z students' hopes for flexible and accessible online learning to replace traditional educational settings [4]. To ensure that education remains relevant and responsive to both current and future demands of many stakeholders, shifts towards teaching with DT are therefore seen as a solution.

The term disruptive learning technologies (DT) refers to an innovation that displaces established technology transforming traditional approaches. The use of DT significantly alters existing ways of learning and teaching and therefore, has a potential to change the current understanding of education. According to the Cambridge dictionary [5], to disrupt means "to prevent something, especially a system, process, or event, from continuing as usual or as expected". Thus, disruption is usually perceived as a negative occurrence triggered by outside factors [6]. Some emerging learning technologies can trigger profound changes and disrupt existing structures and norms, while others do not.

Developing and implementing practice-based e-learning is not just about adopting DT. It is a complex process and also includes changing of organizational culture, processes, and experiences to leverage these technologies effectively. Maturity models are frameworks which outline the key components that in their interconnected roles, are critical factors in realizing maturity objectives. Maturity models can therefore be used as tools to describe and diagnose which improvements should be introduced into processes to achieve the desired change [7]. Analyzing previous work on digital transformation and digital maturity, we define four dimensions to focus on when describing HEIs' maturity to facilitate practice-based e-learning with DT: infrastructure, leadership, community level, and personal level. Each of these dimensions is described by a set of components.

The technological infrastructure within HEIs serves as the backbone for integrating DT into educational practices. Adequate availability of infrastructure and devices supports the quality of the educational process, ensures the seamless operation of educators, students, and administration staff and also guarantees the preservation and protection of data and information [8, 9]. During the Covid-19 pandemic, connectivity to conduct teaching and learning online was proven to be a crucial but also often a missing prerequisite for digital learning [9–16].

Teaching with DT and METAVERSE elements requires enhanced connectivity (minimum 300 Mb network) and more powerful devices (minimum: Intel Core i7 Processor; RAM 16 GB; Graphics Card: RTX 2060), necessitating readiness from both educators' computers and students' devices. Leadership in the digital transformation of HEIs is based on a strategic, holistic approach that encompasses governance, policy alignment, knowledge management, institutional support, and professional development. This comprehensive approach ensures that digital transformation efforts are effective, sustainable, and aligned with the evolving needs of the HEI landscape and supports participants coping with change. One characteristic feature of HEIs is the autonomy of the educators, which allows them to choose the e-learning environments and tools suitable for conducting their teaching and to seize opportunities to experiment with innovative teaching methods using DT.

The community dimension incorporates the components that describe the collective development of transition to integrate DT into teaching practices. The models of digital maturity [17, 18] emphasize the need for strengthening the digital competences by offering professional development activities (e.g., training, observation, co-teaching, becoming a trainer of educators) for educators. In addition, active participation in communities of practice and networking with industry, public sector, NGOs, or startup sector partners serve to share best practices and offer opportunities for knowledge transfer, development, and research of new teaching practices. Quaicoe et al. [19] advocate that educators should be seen as change agents in the digital transition process and, therefore, there is a need to promote cooperation between educators and decision-makers to encourage bottom-up innovation. Personal attitudes of the educators play a significant role in integration of new technologies in higher education. This study explored: What gaps exist in HEIs' use of disruptive learning technologies (DT) for practical e-learning? How is the maturity to conduct remote practical learning with DT reflected by experts, teachers, and students?

2 Methodology

To describe the gaps in the current state of HEIs' maturity to facilitate practice-based e-learning with DT, an online survey was conducted within the Horizon Europe project eDiploma. The survey incorporated questions for four maturity dimensions: infrastructure, leadership, community, and personal dimension. Each dimension encompassed various components (see Table 1). To capture the multifaceted perspectives, the survey was designed to gather responses from different involved parties, including educators (L), students (S), and experts such as technical and didactic support personnel: educational technologists, IT or technical support specialists, faculty members responsible for technology training, and digital policy administration specialists (E). The questions asked of experts incorporated information about the institution (e.g., "There is access to the simulation facilities for lessons in the institution"), while lecturers and students evaluated their own experiences and lived practice (e.g., "I can use the simulation facilities for lessons at my institution"). Statements were evaluated using a 5-point Likert scale, and the respondents had the following options to choose from: Strongly disagree - 1; Somewhat disagree - 2; Neither agree nor disagree - 3; Somewhat agree - 4; Strongly agree - 5. There was also the option to opt-out by choosing Do not know - 0.

A non-probabilistic sampling procedure aiming for a proportional sample was used to gather data from five project partner countries: Spain, Estonia, Bulgaria, Hungary, and Cyprus. The survey was conducted in the local language. The inclusion criteria for the institution were that the entity must be a HEI of the partner country and the desired proportion of the sample was to gather 3–5 responses from experts, 10 responses from educators, and 20 responses from students from each institution.

Aligned with each participant country's Ethics Committee, the survey was piloted from February 2023 to April 2023. Aggregated data from the voluntary respondents from four partner countries' HEIs was merged into a dataset, which included a total of 847 responses from 87 technical and didactic support personnel (E), 327 educators (L), and 433 students (S) from 92 tertiary education institutions.

Reliability for each component of the maturity dimension was verified with Cronbach's alpha, which is a statistic representing internal consistency among items in a dataset. Next, the average score (mean) of each component of the four dimensions was calculated. The results are represented in Table 1. To measure the significant statistical differences, a t-test was conducted. The results are presented in Table 2. Analyses were performed with the statistical software SPSS, version 29.02.

3   Findings

To identify the gaps in HEIs' maturity to facilitate practice-based e-learning with DT, multifaceted perspectives were gathered. For infrastructure, leadership, and community dimensions, responses from experts and lecturers were collected. Experts evaluated the components from an institutional perspective, and lecturers evaluated their own experiences and lived practices. At the community level, the components of traditional e-learning and new e-learning were evaluated by educators and students, as they are participating in the teaching and learning activities. The personal level dimension collected evaluations from experts, educators, and students (except for the component 'effort to change').

3.1   Infrastructure Dimension

The evaluation of the current state of the infrastructure dimension (see Table 1) showed a moderately positive assessment (between neither agree/disagree (3) and somewhat agree (4)) for all components by both experts and educators. Experts were hesitant or somewhat agreed about the availability of software to facilitate online learning (e.g., centrally provided LMS, LDS, video conferencing tools) (M = 3.61, SD = 1.08). For the internet connectivity and processor adequacy to support METAVERSE development and experience, experts and educators reported hesitantly or tended to somewhat agree (M = 3.47, SD = 1.24; M = 3.48, SD = 1.13). The availability, storage capacity, and compatibility of centrally selected repositories to manage digital learning resources were evaluated by experts and educators somewhat positively (M = 3.73, SD = 0.85; M = 3.63, SD = 1.06). The experts and educators were rather hesitant about the availability of learning analytics (M = 3.39, SD = 1.16; M = 3.26, SD = 1.10).

There was a significant difference between experts' and educators' views regarding the availability of physical rooms to facilitate distance learning (see Table 2), including teaching with DT (p < .001). This indicates a lack of practical facilities.

Table 1. Maturity to Conduct Practice-Based E-Learning with Disruptive Learning Technologies in HEIs.

	Component	Experts				Lecturers				Students			
		N	M	SD	α	N	M	SD	α	N	M	SD	α
Infrastructure	Software	87	3.61	1.08	0.78								
	Physical classrooms	86	3.75	0.84	0.82	327	3.37	0.97	0.85				
	Connections & Wi-Fi	82	3.47	1.24	0.84	323	3.48	1.13	0.83				
	Digital assets & repositories	85	3.73	0.85	0.76	317	3.63	1.06	0.85				
	Data & Analytics	85	3.39	1.16	0.82	320	3.26	1.10	.80				
Leadership	Regulations & Policies	86	3.75	0.78	0.91								
	Roles & Responsibilities	86	3.80	0.96	0.86								
	Lecturer's autonomy	86	3.77	0.75	0.73	327	3.61	0.83	0.77				
	Lecturer's professional development plan	84	3.54	1.07	0.83	324	3.39	1.05	0.73				
Community	Support	85	3.66	0.99	0.87	324	3.17	1.04	0.85				
	Collaboration within the organization	86	3.61	0.86	0.78	327	3.04	1.03	0.82				
	Collaboration beyond the organization	85	3.69	0.88	0.71	322	2.87	1.21	0.85				
	Participation in training	84	3.66	1.01	0.82	322	3.19	1.07	0.93				
Personal	Traditional e-learning					323	3.08	1.09	0.86	430	3.39	0.83	0.78
	New e-learning					321	2.88	1.18	0.91	428	2.74	1.13	0.79
	Competences (L)	81	3.49	1.12	0.76	321	3.12	1.11	0.94				
	Competences (S)	80	3.59	1.02	0.79	297	3.11	1.11	0.74	413	3.29	1.20	0.94
	Effort to change	81	3.84	0.84	0.76	312	3.55	0.91	0.70				
	Risk evaluation	83	4.09	0.87	0.89	315	3.51	1.02	0.89	416	3.37	1.10	0.87
	Impact	83	3.86	0.74	0.92	319	3.69	0.78	0.91	417	3.62	0.79	0.89

3.2 Leadership Dimension

Experts' results (see Table 1) indicated that HEIs' strategic and regulatory preparation for teaching with new learning technologies is somewhat agreed upon (M = 3.75, SD = 0.78). Furthermore, experts somewhat agreed (M = 3.80, SD = 0.96) that HEIs have established institutional roles and responsibilities, such as designated departments responsible for coordinating the development of digital infrastructure and educational tools, training and mentoring for digital transformation, and providing technological-didactical support for lecturers in digitally mediated practices. Experts and educators were still hesitant about the availability of plans for developing digital competences and initiatives to motivate the development of online courses and teaching approaches with innovative technologies in HEIs (M = 3.54, SD = 1.07; M = 3.39, SD = 1.05). Educators somewhat agreed/disagreed that they would need support while developing and facilitating e-learning courses (M = 3.17, SD = 1.04), compared to the readiness of HEIs to offer the support, which was evaluated higher by experts (M = 3.66, SD = 1.04). Both experts (M = 3.77, SD = 0.75) and educators (M = 3.61, SD = 0.81) somewhat agreed that the learning process planning is flexible enough and supports educators in choosing different digital tools for building their courses, testing out new technologies for learning, using repositories of their own choice, and seizing opportunities, if available, to teach with new technologies.

3.3 Community Dimension

The collaboration component (see Table 1) estimates the level of reciprocal knowledge and experience exchange regarding new, innovative technologies, both within the institution and with external entities. Experts evaluated the institutional collaborative setups, whereas educators reflected on their engagement with these arrangements. There was a significant difference (see Tables 1 and 2) in experts' and educators' evaluations (M = 3.61, SD = 0.86; M = 3.04, SD = 1.03) about the opportunities created for educators' engagement internally, in institutional settings for best practices sharing and involvement in digital agendas and strategy development (p < 0.01). The same trend, significant difference, appeared in experts' and educators' assessments of participating in networks with key stakeholders, including industry, public sector, NGOs, startups, and alumni (M = 3.68, SD = 0.88; M = 2.87, SD = 1.21; p < 0.01).

The professional development of digital competencies was manifested in diverse learning opportunities: through independent self-improvement (e.g., developing technologies), participating in training, sharing experiences with colleagues in communities of practice, participating in classroom observations, networking with partners (visiting workplaces), and conducting professional training for colleagues. Educators rated their actual participation in digital competencies development significantly different from experts, who evaluated the opportunities offered institutionally for participation in the professional development of digital competencies (M = 3.66, SD = 1.01; M = 3.19, SD = 1.07; p < 0.01). Educators and students estimated the teaching and learning practice components focused on distance learning practices in HEIs. For educators, traditional distance learning practices were embedded with rather somewhat agreeing statements (see Table 1) for conducting learning courses in synchronous, asynchronous, blended, and/or hybrid/flexible forms; using different group work methods during teaching in a distant mode, as well as conducting collaborative distance learning in cooperation with colleagues, students, or external partners. The new, innovative distant learning component focused on distant learning employing DT (AI, AR, VR, robots, chatbots, virtual games). Students evaluated the same statements from the perspective of actual participation in traditional distance learning and new learning practices.

There was a significant difference in educators' and students' evaluations about traditional e-learning (p < 0.01). Both students and educators had little experience with teaching and learning with DT: educators had a mean of 2.88 (SD = 1.18) and students had a mean of 2.74 (SD = 1.13).

3.4 Personal Dimension

Under the competence dimension, competencies for developing digital learning scenarios and resources with DT (VR, AR, AI, robots, etc.) and personalizing learning, including adapting e-learning situations for learners with special needs along with the knowledge about the potential of disruptive technologies for humans, for learning, and for sustainability were assessed by educators and experts (see Table 1). Experts gave evaluations based on their beliefs about the educators, while educators self-evaluated their own competences. There was a significant difference between educators' self-evaluation of their competences and experts' beliefs about the educators' competences (M = 3.12, SD = 1.11; M = 3.49, SD = 1.12; p = 0.003).

Student competencies to participate in practical online courses using DT and the knowledge about the pros and cons of using DT in learning to make justified decisions about their learning choices were assessed by experts and educators, while students evaluated their own competences. The results revealed that there is a significant difference between the experts' and educators' evaluations of the students' competences (M = 3.59, SD = 1.02; M = 3.11, SD = 1.11; p < 0.01) and educators' and students' evaluations of student competences (M = 3.11, SD = 1.11; M = 3.29, SD = 1.20; p = 0.05). There was a significant difference in the scores of experts compared to educators (M = 3.84, SD = 0.84; M = 3.55, SD = 0.91; p = 0.010) regarding the effort (requirement of resources, staff training, new regulations, and norms) to introduce DT in classes (see Tables 1 and 2).

In terms of risk evaluation, experts believe that institutions must evaluate the potentials and threats of DT to learning, sustainability, and health and wellbeing aspects (M = 4.07, SD = 0.87). Educators evaluated their current practice of considering the potentials and threats of using DT for humans, learning potentials, sustainability, and wellbeing as agreeing rather than disagreeing (M = 3.51, SD = 1.02).

Table 2. Significant Differences in Experts' (E), Educators' (L), and Students' (S) Perspectives on the Digital Maturity Components in HEIs.

Component	Group	N	mean	df	t-value	Sig. (t-tailed)
Physical classrooms	E	86	3.75	411	3.32	<0.01
	L	327	3.37	149.94		
Support	E	85	3.66	407	3.09	<0.01
	L	342	3.17	137.08		
Collaboration within the organization	E	86	3.61	411	5.27	<0.01
	L	327	3.04	155.47		
Collaboration beyond the organization	E	85	3.69	401	7.10	<0.01
	L	322	2.87	177.77		
Participation in training	E	84	3.66	404	3.64	<0.01
	L	322	3.19	135.21		
Traditional e-learning	L	323	3.08	751	-4.52	<0.01
	S	430	3.39	580.66		
Competences of lecturers	E	81	3.49	400	2.7	0.003
	L	321	3.12	122.45		
Competences of students	E	80	3.59	375	3.49	<0.01
	L	297	3.11	133.39		
	L	297	3.11	708	-1.96	0.05
	S	413	3.29	666.64		
Effort to change	E	81	3.84	391	2.57	0.010
	L	312	3.55	132.20		
	L	312	3.55	719	3.53	<0.01
	S	409	3.30	702.24		
Risk evaluation	E	83	4.09	396	4.74	<0.01
	L	315	3.51	132.88		
	E	83	4.09	497	5.68	<0.01
	S	416	3.37	140.02		
Impact	E	83	3.86	4.98	2.70	0.08
	S	417	3.62	121.25		

The findings showed that experts ($M = 3.86$, $SD = 0.74$), educators ($M = 3.69$, $SD = 0.78$), and students ($M = 3.62$, $SD = 0.79$) all have somewhat positive attitudes that integrating DT into education strengthens the sector's resilience, brings additional value to the learning experience, advances the social and collaborative dimension of learning, promotes students' learning results, and advances human abilities. Furthermore, they somewhat agree that employing disruptive technologies in education enhances students' skills for digital jobs, promotes ecosystem sustainability, and is more cost-effective compared to traditional face-to-face learning. There is a significant difference between the experts' and students' attitudes towards the impact of using disruptive learning technologies ($M = 3.86$, $SD = 0.74$; $M = 3.62$, $SD = 0.79$; $p = 0.08$) (see Tables 1 and 2).

4   Discussion

The findings provide insight into what gaps exist in tertiary education's use of DT for practice-based e-learning, and how the maturity to conduct remote practical learning with DT is reflected by HEI experts, teachers, and students.

In addressing gaps in HEIs' use of DT in practice-based e-learning, it is essential to recognize the complex interrelations between the infrastructure, leadership, community, and personal level components. Educators' and students' evaluations revealed that the actual experience in teaching and learning with DT in HEIs is limited. This could be explained by the infrastructure not being available for such teaching yet in HEIs. Infrastructure can be seen as the prerequisite for teaching and learning with new learning technologies. In the transition to teaching with DT, institutions must upgrade the current technology park and adjust the physical room performance to support distance learning. Lecturers' lower ratings of the readiness and equipment of the institution's physical lecturing rooms to conduct digital learning practices suggest potential hesitancy or difficulty in actual accessibility, adoption, and integration of digital technologies into their teaching practices. This could lead to underutilization of available technologies and limit the effectiveness of digital learning practices. Also, there is a concern in teaching with DT related to connectivity and processor adequacy. To participate in the distance learning process, reliable and high-capacity internet connections and devices are needed from educators but also from students, to ensure seamless integration. Studies conducted during COVID-19 [9–16] showed that not all participants had sufficient devices and network connections for distance learning. In teaching with disruptive technologies, these resources become even more critical as prerequisites for participation in learning activities.

Even if there is limited actual experience in facilitating or participating in learning activities with DT, and although the risks and efforts involved in shifting towards teaching and learning with new learning technologies are evaluated with the highest mean scores, there is still a positive attitude in all participant groups towards the positive impact of engaging DT from both learning and sustainability perspectives. The techno-positive attitude could stem from the influence of industry leaders, policymakers, and the expectations of digital native generations, all of whom champion the benefits of technology.

One of the gaps in tertiary education institutions' maturity to facilitate remote practical learning with DT is the gap between the view of the institution's actions towards the transition and the actual lived experience of the educators. There are trends showing that from a leadership perspective, the options for participating in professional development, communities of practice, and networking are initiated, but according to the educators' self-evaluation, there is still room for the utilization of those activities in lived practice. This highlights the need to broaden the collaboration between experts and educators within organizational settings and to expand cooperation with external partners, as well as empowering educators by supporting more active participation in personal and professional development.

Acknowledgments. Research supported by the e-DIPLOMA project, number 101061424, funded by the European Union. Views and opinions expressed are, however, those of the authors only and do not necessarily reflect those of the European Union or the European Research Executive Agency (REA). Neither the European Union nor the granting authority can be held responsible for them.

References

1. Goulart, V.G., Liboni, L.B., Cezarino, L.O.: Balancing skills in the digital transformation era: the future of jobs and the role of higher education. Ind. High. Educ. **36**(2), 118–127 (2021)
2. Grigorescu, A., Zamfir, A., Sigurdarson, H.T., Carlson, E.L.: Skill needs among european workers in knowledge production and transfer occupations. Electronics **11**(18) (2022)
3. European Commission: Digital Education Action Plan 2021–2027. Resetting education and training for the digital age. https://education.ec.europa.eu/focus-topics/digital-education/action-plan. Accessed 12 March 2024
4. Mospan, N.: Trends in emergency higher education digital transformation during the COVID-19 pandemic. J. Univ. Teach. Learn. Pract. **20**(1), 50–70 (2023)
5. Cambridge University Press: Disruptive. In Cambridge Dictionary. https://dictionary.cambridge.org/us/dictionary/english/disruptive. Accessed 27 Feb 2024
6. Boucher, P., Bentzen, N., Laţici, T., Madiega, T., Schmertzing, L., Szczepański, M.: Disruption by Technology: Impacts on Politics, Economics and Society. European Parliamentary Research Service. https://www.europarl.europa.eu/stoa/en/document/EPRS_IDA(2020)652079. Accessed 10 March 2024
7. Santos-Neto, J.B.S.D., Costa, A.P.C.S.: Enterprise maturity models: a systematic literature review. Enterprise Inform. Syst. **13**(5), 719–769 (2019)
8. Al-Malah, D.K.A., Majeed, B.H., ALRikabi, H.T.S.: Enhancement of educational technology by using 5G networks. Int. J. Emerg. Technol. Learn. **18**(1), 137–151 (2023)
9. Jain, N., Thomas, A., Gupta, V., Ossorio, M., Porcheddu, D.: Stimulating CSR learning collaboration by the mentor universities with digital tools and technologies: an empirical study during the COVID-19 pandemic. Manag. Decis. **60**(10), 2824–2848 (2022)
10. Villarica, M.V.: The effectiveness of flipped classrooms in distance education during the COVID–19 pandemic. Int. J. Comput. Sci. Res. **7**, 2296–2314 (2023)
11. Khan, M.A., Ahmed, S.K., Khan, A.Z.: Issues and challenges faced by higher education institutions in the e-learning modes: a systematic literature review. Bahria J. Profession. Psychol. **21**(1), 58–72 (2022)
12. Adedoyin, O.B., Soykan, E.: Covid-19 pandemic and online learning: the challenges and opportunities. Interact. Learn. Environ. **31**(2), 863–875(2020)

13. Wang, C., Cheng, Z., Yue, X., McAleer, M.: Risk management of COVID-19 by universities in China. J. Risk Finan. Manage. **13**(2), 36 (2020)
14. Bao, W.: COVID-19 and online teaching in higher education: a case study of Peking University. Hum. Behav. Emerg. Technol. **2**(2), 113–115 (2020)
15. Doyumğaç, I., Tanhan, A., Kıymaz, M.S.: Understanding the most important facilitators and barriers for online education during COVID-19 through online photovoice methodology. Int. J. High. Educ. **10**(1), 166 (2020)
16. Marcial, D.E., Villariza, C.R.R., Launer, M.A., Binarao, S.M.: Technology integration in the workplace: a global study. Webology **19**(3) (2022)
17. Costa, P., Castaño-Muñoz, J., Kampylis, P.: Capturing schools' digital capacity: Psychometric analyses of the SELFIE self-reflection tool. Comput. Educ. **162**, 104080 (2021)
18. Rossmann, A.: Digital maturity: conceptualization and measurement model. ResearchGate (2019)
19. Quaicoe, J.S., Ogunyemi, A., Bauters, M.: School-based digital innovation challenges and way forward conversations about digital transformation in education. Educ. Sci. **13**(4), 344 (2023)

Factors Affecting the System Usability of a Maritime Learning Analytics Dashboard prototype

Helene Krabbel[1(✉)], Ziaul Haque Munim[1(✉)], Morten Bustgaard[1], and Emilia Lindroos[2]

[1] Faculty of Technology, Natural and Maritime Sciences, University of South-Eastern Norway, Horten, Norway
s.krami99@gmail.com, ziaul.h.munim@usn.no
[2] Faculty of Technology and Seafaring, Novia University of Applied Sciences, Turku, Finland

Abstract. To successfully implement a Learning Analytics Dashboard (LAD), the perceived usability by its users' needs to be assessed at different stages of development. This study examines how the demographic background of students in maritime education and training, their academic experiences, and perception toward artificial intelligence (AI), data security etc. influences the perceived usability of a LAD prototype. Two LAD visualizations under development have been shared with two groups of potential users: (1) line graphs based on desktop simulator log data, and (2) heat maps based on eye-tracking data. A follow up survey was conducted incorporating the System Usability Scale (SUS) and focusing on related qualitative factors. A total of 63 useable responses were analysed using machine learning (ML). The findings reveal that users who were exposed to line graphs have a higher perceived usability than those exposed to eye-tracking heat maps. Further, those with a higher degree of trust in Artificial Intelligence (AI) also have a higher perceived system usability.

Keywords: Learning Analytics · System Usability Scale · Machine Learning · Eye tracking

1 Introduction

The technological advancements in recent years together with an increasingly digitalized learning environment have created large data sets of student's behaviour in academic learning [1, 2]. Learning Analytics (LA) is allowing both teachers and students to better understand this learning behaviour and the learning process by analysing this data [2]. Maritime Learning Analytics incorporates the principles of LA into Maritime Education and Training (MET). As assessment in maritime simulator training is still mainly relying on instructor feedback, research for alternative assessment approaches has gained popularity over the last years [3]. LA can support the implementation of objective assessment based on for example simulator log data.

T. E. Kim et al. (Eds.): MIS4TEL 2024, LNNS 1274, pp. 11–22, 2025.
https://doi.org/10.1007/978-3-031-84170-5_2

Once the data is collected and processed, it can be presented with the help of a Learning Analytics Dashboard (LAD). Descriptive LADs can provide comprehensive feedback to the target audience by displaying information such as performance and activity, while predictive LADs can also predict information on future performance outcomes or drop-out rates [4, 5]. However, to unlock the full potential of LADs, they need to cater to their users' needs, prompting scholars to investigate the factors influencing the perceived usability of LADs [2, 6].

Figure 1 shows the LAD prototype that was presented to the students. It provides an overview of relevant navigation cues (i.e. speed, course, distance to all ships, heading, main propeller revolutions and the main rudder angle) over time. The input data stem from the simulator log data and is visualized as line graphs. Students can choose between their personal overview and a comparative display. Figure 2 shows two different heat map examples, on the left is the visualization of a novice seafarer while on the right the heat map of an experienced seafarer can be seen.

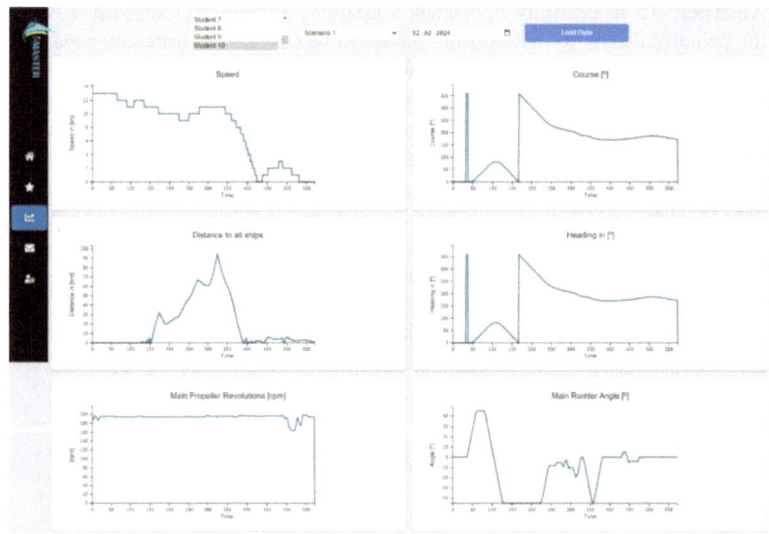

Fig. 1. Visualization from the i-MASTER Prototype (single student example) based on simulator log data of a Williamson turn exercise conducted in February 2024 at USN, taken from, taken from [7]

To evaluate the usability of products and services such as LADs, the System Usability Scale (SUS) provides a valuable tool. Developed by John Brooke in 1995 it is widely used to measure both usability and user satisfaction [9]. To determine the SUS Score, in the original version 10 questions are posed alternatingly between positive and negative as a five-point Likert scale, ranging from strongly disagree to strongly agree. The SUS has not only proven itself in general science, it has also been used successfully in numerous instances in educational technology [10, 11].

Jo and Park [12] found that the use of a LAD did not have a significant impact on students' learning achievements, and they are not the only ones, see for instance [13].

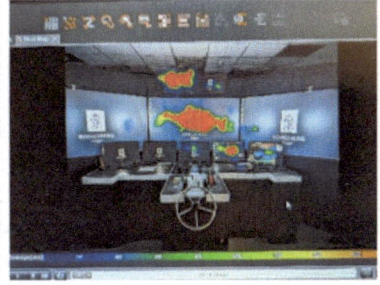

Fig. 2. Heatmap visualization of a novice seafarer (left) and an experienced seafarer (right), taken from [8]

However, visualized information has been found to impact the level of understanding, while a correlation between the students' satisfaction with the LAD and behavioural changes as well as level of understanding has been observed [12]. Hence, to unlock the full potential of LADs, the perceived system usability plays an important role. Yet research is lacking, especially in the field of maritime education and training.

The following research question will guide this study: How do demographic factors and academic training of maritime students influence their perception of the usability of a maritime Learning Analytics Dashboard prototype? For the remainder of this study, first the data collection process is described, where data from a structured web-survey and in-person interviews are collected. To analyse the results, machine learning (ML) is used to identify the most relevant factors affecting the system usability of a maritime learning analytics dashboard prototype. The results are presented in Sect. 3 of this study. The study is concluded in the last section.

2   Data and Methodology

2.1   Data Collection

To collect data on factors affecting the system usability of a maritime LAD prototype, an online survey was designed. The first part contains multiple statements and questions based on the System Usability Scale by [9]. The version adapted in this study uses only positively formed statements in a five-point Likert scale [10]. Hence, to calculate the comprehensive SUS Score, the sum of each respondent's rating for the ten questions is multiplied by two to get a score between 0 and 100.

The second part of the survey is structured as open-ended interview-like questions, which were developed based on [14]. The questions targeted the student's personal background (i.e., gender, current study level), their familiarity with data visualizations, and their first impression when presented the prototype. To allow for a more nuanced impression reflection, the students were asked if they found anything surprising, how the LAD might facilitate their performance, and whether it is a suitable tool for their study program. The open-ended questions concluded aiming at potential challenges in the implementation including concerns regarding student identity, data security and AI ethics.

The data have been collected in February 2024 from two Nordic maritime universities. The participation in the survey was voluntary, and consent has been taken from each participant. A total of 64 students participated. Data of one participant were disregarded due to straight lining, leaving a total of 63 answers for the analysis. The collected data were the perceptions of either line graphs based on desktop simulator log data or heat maps from eye-tracking data from a full-mission simulator exercise. Additionally, also in February 2024, three students from a third maritime university were interviewed on their campus. The in-person interviews were focused on eye-tracking learning analytics. The data collection process can be seen in Fig. 3.

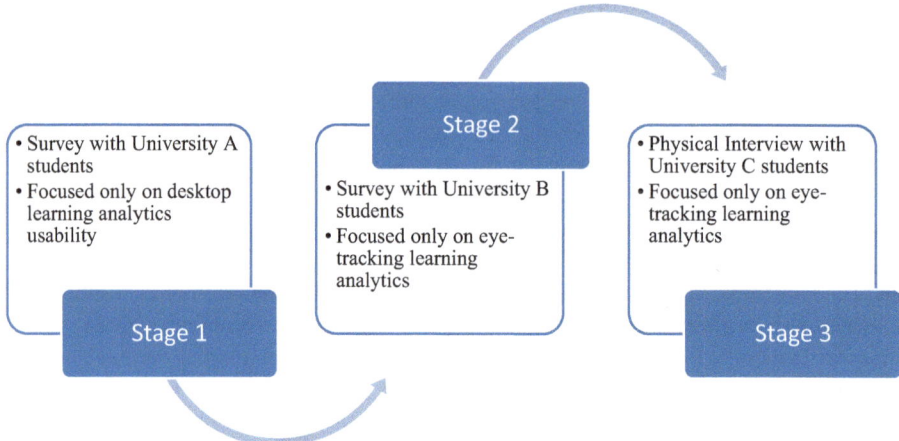

Fig. 3. Data Collection Process

The interview questions were coded into a maximum of five variables per question. In case of missing answers, "missing" was used. The coded interview answers together with the calculated SUS Score were formatted in a table which then was uploaded to DataRobot to determine if and to what extent each input feature influences the SUS score.

2.2 Respondents' Background

The background overview of the 63 respondents is presented in Table 1. Notably, there were fewer female participants which can be attributed to seafaring still being a male-dominated profession. The majority of the participants is in their 2nd year of studies and already has experienced full-mission simulator training. 38 participants are from University A who were demonstrated the line graph-based LAD and 25 from University B who were exposed to eye tracking heat map visualization.

2.3 Automated Machine Learning

With latest development in Machine Learning (ML), Automated Machine Learning (AutoML) allows to run a large number of algorithms within a limited time period and

Table 1. Respondents' background

Variable	Categories	Frequency	Percent	Valid Percent	Cumulative Percent
Gender	Female	8	12,7	12,7	12,7
	Male	55	87,3	87,3	100,0
	Total	**63**	**100,0**	**100,0**	
Current study level	1st year student	23	36,5	36,5	36,5
	2nd year student	36	57,1	57,1	93,7
	3rd year student	4	6,3	6,3	100,0
	Total	**63**	**100,0**	**100,0**	
Experience with full-mission simulators	0 = No	28	44,4	44,4	44,4
	1 = Yes	35	55,6	55,6	100,0
	Total	**63**	**100,0**	**100,0**	
Institution	A/Desktop exercise	38	60,3	60,3	60,3
	B/Full-mission exercise	25	39,7	39,7	100,0
	Total	**63**	**100,0**	**100,0**	

suggesting the best performing model [15]. Furthermore, it eradicates the need for human intervention when performing the repetitive tasks in the ML life cycle [15, 16].

AutoML Platforms

To analyse the data collected in this study the AutoML platform DataRobot was used. DataRobot's AutoML tool was launched in 2015, enabling users to make use of different ML tasks and models with limited previous experience with coding and data analytics [15, 17]. Besides DataRobot, numerous other AutoML platforms exist, such as Tensor-flow, Keras, Auto-Weka, Auto-ml, H2O-Automl and Google Cloud Automl amongst others [15].

Input Features

To determine whether demographic, background, familiarity with visualizations, concerns about AI, and concerns about identity and data identity concerns affect the SUS Score, several models were trained using AutoML. The calculated SUS Score has been selected as the target (i.e. dependent) variable. Table 2 shows the properties of each input feature. In addition to demographic variables, several input features stem from the data collection survey's open-ended interview-like questions. These features were coded manually for each respondent based on the text responses. As reported in Table 2, these variables were treated as categorical, with unique column presenting the number of categories. Experience with full mission simulators and eye-tracking were coded as dummy variables, where the numerical values of 0 (no) and 1 (yes) were used. In the model training setup in DataRobot 3-fold cross-validation and 30% holdout sample were used.

Table 2. Data properties of features

Feature Name	Index	Var Type	Unique	Missing
Gender	1	Categorical	2	0
Current study level	2	Categorical	3	0
Experience with full_mission simulators	3	Numeric	2	0
Familiar with visualization of data	4	Categorical	5	0
First impression_coded	5	Categorical	4	0
Surprise_coded	6	Categorical	4	0
Facilitate performance coded	7	Categorical	5	0
Identity concern_coded	8	Categorical	4	0
Suitability coded	9	Categorical	4	0
Data security concerns_coded	10	Categorical	3	0
AI ethics concerns_coded	11	Categorical	5	0
Eye-tracking	12	Numeric	2	0
SUS Score	TARGET	Numeric	21	0

Model Evaluation

The R-squared was used as ML model performance optimization metric, as it measures the proportion of the total variation of the outcomes explained by the model. 74 models were trained by DataRobot, where the best performing model was Light Gradient Boosting on ElasticNet Predictions with 22,59% validation and 30,81% cross-validation R-squared. The five best performing models are reported in Table 3. Note that four out of five are Light Gradient Boosting on ElasticNet Predictions with varying sample sizes and properties.

Table 3. Best performing models (sorted by CV accuracy)

No.	Model	Sample size	Validation	Cross-Validation	Holdout
1	Light Gradient Boosting on ElasticNet Predictions	100%	0.2259	0.3081	0.2991
2	Light Gradient Boosting on ElasticNet Predictions	46%	0.3141	0.2911	0.2888

(*continued*)

Table 3. (*continued*)

No.	Model	Sample size	Validation	Cross-Validation	Holdout
3	Auto-tuned K-Nearest Neighbors Regressor (Euclidean Distance)	46%	0.3485	0.2507	0.3309
4	Light Gradient Boosting on ElasticNet Predictions (reduced feature list)	46%	0.1877	0.2451	0.2681
5	Light Gradient Boosting on ElasticNet Predictions	70%	0.1417	0.2111	0.2837

3   Results

3.1   Feature Association

The feature association gives insights into the degree of association between the features used in the analysis. Figure 3 displays the feature association matrix. The three clusters in red, blue, and green have a high correlation between themselves. The darker the colour, the better the association. The strongest association can be witnessed between the study level and whether full mission experience exists (+0,654). The 2nd highest association is between data security and suitability (+0,395).

3.2   Factors Influencing SUS-Score

The feature impact in Fig. 4 depicts the relative influence of each input feature on the target variable, the calculated SUS score. The feature with the highest impact is exposure to heat map-based learning analytics from eye-tracking data. Hence, it is depicted with 100%. All other feature impacts are shown relative to the eye-tracking feature, which is followed by AI ethics concerns and familiarity with data visualization.

3.3   Predicted SUS-Scores for Each Factor

The feature effects depict how variations in each feature's value influence the predictions made by the model for the target variable, that is, the SUS Score. The partial dependence shows how the target feature predictions are influenced when one input feature's value is changed, while all other values remain constant. Figure 5 shows the feature effect for each variable. Users who viewed line graphs reported greater perceived usability compared to those who were exposed to eye-tracking heat maps (Fig. 6a). Additionally, individuals with a higher level of trust in Artificial Intelligence (AI) also experienced greater perceived usability of the system (Fig. 6b). Those who are neutral regarding their familiarity with data visualization are relatively unlikely to perceive the LAD prototype useful (Fig. 6c). Those who responded that LAD facilitates their performance significantly also perceive higher usability (Fig. 6d). The second year students and those who were surprised after seeing the LAD visualizations also perceive higher usability (Fig. 6e and f). Similarly, Fig. 6g–i can be interpreted.

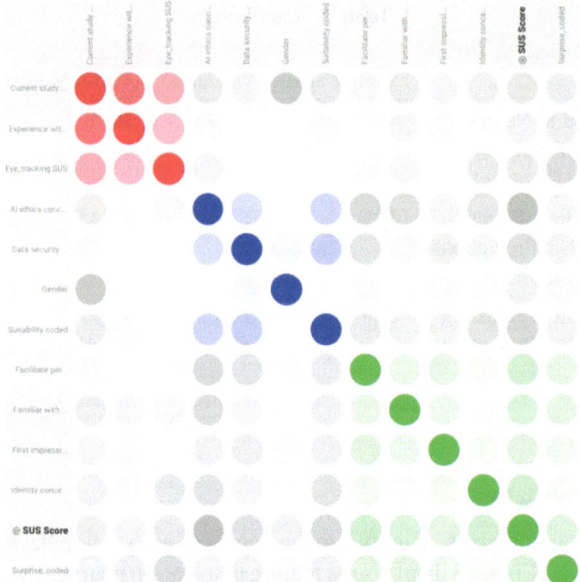

Fig. 4. Feature Association Matrix

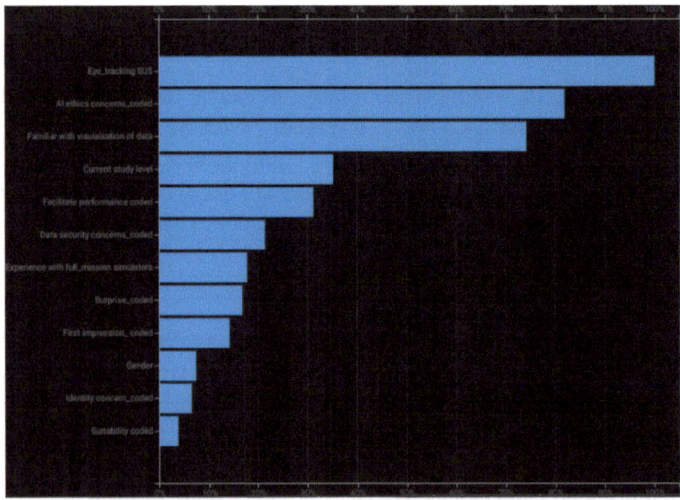

Fig. 5. Feature Impact

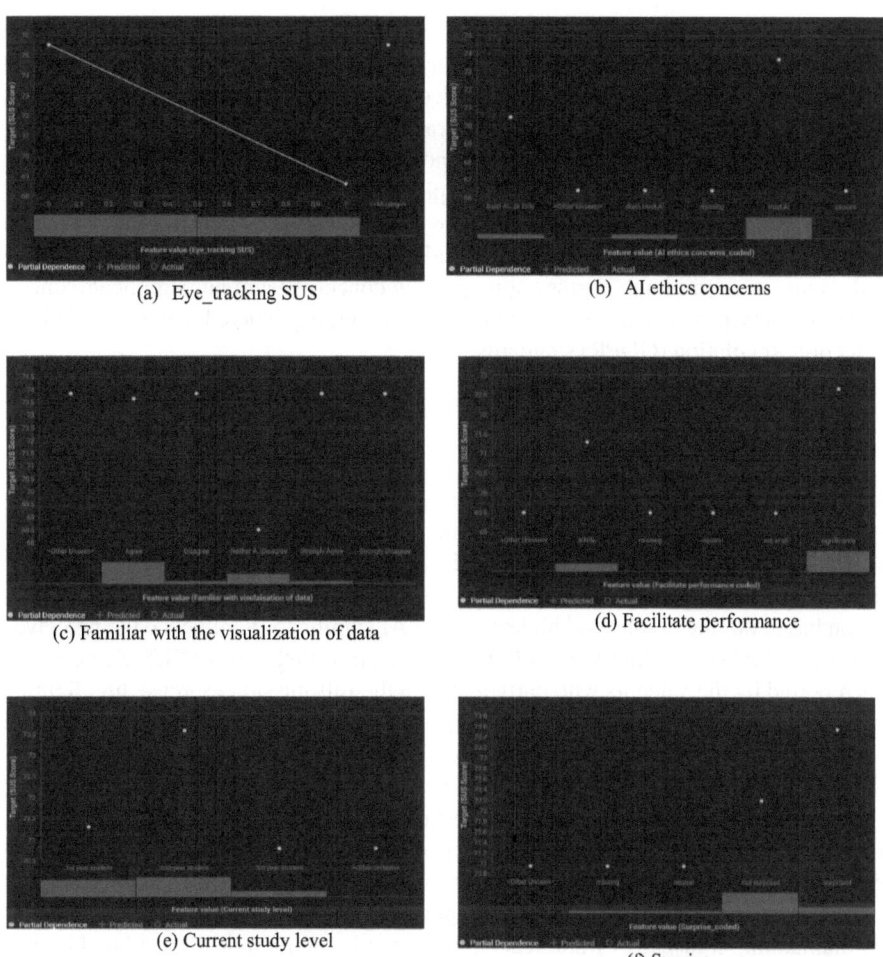

(a) Eye_tracking SUS

(b) AI ethics concerns

(c) Familiar with the visualization of data

(d) Facilitate performance

(e) Current study level

(f) Suprise

Fig. 6. Feature Effects on SUS ScoreConclusions

3.4   Findings from the Interviews

Based on the campus interviews with three students who participated in the eye-tracking experiment in the full-mission simulation exercise, further support and interpretation of the findings can be derived. The students reported finding it beneficial to view their own eye-tracking heat maps, which enhanced their curiosity and were perceived as useful for reflecting on their learning processes. The utility of these heat maps appears to be influenced by the difficulty of tasks and the students' prior experiences with the content, suggesting that familiarity with the context may affect heat map results. Further, heat maps were particularly valued for reviewing completed training tasks, as they clearly indicated any aspects or required actions that were missed.

However, while eye-tracking heat maps are helpful, they may not be the most effective standalone method for evaluating performance due to the variability in successful strategies across different tasks. Thus, they are better used as complementary tools in assessment frameworks. One technical challenge noted was the issue of movement in dynamic settings, such as during collaborative tasks on a bridge simulation, which can cause inaccuracies in data captured by eye-tracking glasses.

Despite these technical issues, there was no opposition to the use of AI in the learning process. Students expected a simplified explanation of the AI algorithms employed, which suggests a need for transparency in the use of such technologies. Concerns about potential negative impacts, such as decreased motivation or complacency from seeing predictions, were minimal. Furthermore, privacy concerns were not significant among the participants, provided that data handling and processing adhered to the General Data Protection Regulation (GDPR) standards.

4   Conclusions

This study examines how students' backgrounds, and several other related factors affect their perceived system usability of a maritime LAD. Data were collected from students in three Nordic maritime institutions. The data were analysed using automated machine learning approach. The findings reveal that specific factors, such as exposure to visualization based on line graphs and higher trust in AI significantly influence the perceived usability of LADs. The interviews offer an explanation to the lower SUS-Score which was reported by the students who participated in the full-mission exercise, highlighting eye tracking data collection as a technical burden. Another explanation may lie in graph literacy, as line graphs tend to be easier to understand and interpret initially.

One of the key findings is that the students who were presented with line graphs reported a higher usability. Furthermore, the AutoML approach allowed to explore which input features had a higher or lower impact on the SUS score. The findings indicate a general acceptance and recognition of the value of integrating learning analytics with the help of AI in educational settings. The importance of addressing technical challenges and maintaining transparency and legal compliance in their application are highlighted too. These insights underscore the necessity of considering user-centric design and ethical considerations in the development of educational technologies to implement the successful use in larger scale.

Acknowledgement. This research is funded by the European Union's Horizon Europe research and innovation programme under grant agreement No 101060107.

References

1. Russell, J.-E., Smith, A., Larsen, R.: Elements of Success: supporting at-risk student resilience through learning analytics. Comput. Educ. **152**, 103890 (2020). https://doi.org/10.1016/j.com pedu.2020.103890
2. Schumacher, C., Ifenthaler, D.: Features students really expect from learning analytics. Comput. Hum. Behav. **78**, 397–407 (2018). https://doi.org/10.1016/j.chb.2017.06.030

3. Munim, Z.H., Krabbel, H., Haavardtun, P., Kim, T.-E., Bustgaard, M., Thorvaldsen, H.: Scenario design, data measurement, and analysis approaches in maritime simulator training: a systematic review. In: Kubincová, Z., Caruso, F., Kim, T., Ivanova, M., Lancia, L., Pellegrino, M.A. (eds.) Methodologies and Intelligent Systems for Technology Enhanced Learning, Workshops - 13th International Conference, LNCS, vol. 769, pp. 39–47. Springer Nature Switzerland, Cham (2023). https://doi.org/10.1007/978-3-031-42134-1_4

4. Peraić, I., Grubišić, A.: Development and evaluation of a learning analytics dashboard for moodle learning management system. In: Meiselwitz , G., et al. (eds.) HCI International 2022 - Late Breaking Papers. Interaction in New Media, Learning and Games, LNCS, vol. 13517, pp. 390–408. Springer Nature Switzerland, Cham (2022). https://doi.org/10.1007/978-3-031-22131-6_30

5. Ramaswami, G., Susnjak, T., Mathrani, A., Umer, R.: Use of predictive analytics within learning analytics dashboards: a review of case studies. Tech Know Learn **28**(3), 959–980 (2023). https://doi.org/10.1007/s10758-022-09613-x

6. Munim, Z.H., et al.: User requirements for learning analytics dashboard in maritime simulator training. In: 2023 IEEE International Conference on Industrial Engineering and Engineering Management (IEEM), pp. 0406–0410. IEEE, Singapore, Singapore (2023). https://doi.org/10.1109/IEEM58616.2023.10406321

7. Kjeldsberg, F., Bustgaard, M., Munim, Z.H., Haavardtun, P., Ujkani, A.: Usability testing and evaluation of the learning analytics for remote maritime simulation, i-MASTER (2024)

8. Kjeldsberg, F., Haavardtun, P., Bhagat, S., Munim, Z.H. Lindroos, E.: Usability testing and evaluation of the learning analytics for on-site maritime simulation', i-MASTER (2024)

9. Brooke, J.: SUS: a quick and dirty usability scale. Usability Eval. Ind. **189** (1995)

10. Lewis, J.R.: The system usability scale: past, present, and future. Int. J. Hum.-Comput. Interact. **34**(7), 577–590 (2018). https://doi.org/10.1080/10447318.2018.1455307

11. Vlachogianni, P., Tselios, N.: Perceived usability evaluation of educational technology using the System Usability Scale (SUS): a systematic review. J. Res. Technol. Educ. **54**(3), 392–409 (2022). https://doi.org/10.1080/15391523.2020.1867938

12. Jo, I.-H., Park, Y.: Development of the Learning Analytics Dashboard to Support Students' Learning Performance' (2015). https://doi.org/10.3217/JUCS-021-01-0110

13. Susnjak, T., Ramaswami, G.S., Mathrani, A.: Learning analytics dashboard: a tool for providing actionable insights to learners. Int. J. Educ. Technol. High. Educ. **19**(1), 12 (2022). https://doi.org/10.1186/s41239-021-00313-7

14. Rets, I., Herodotou, C., Bayer, V., Hlosta, M., Rienties, B.: Exploring critical factors of the perceived usefulness of a learning analytics dashboard for distance university students. Int. J. Educ. Technol. High. Educ. **18**(1), 46 (2021). https://doi.org/10.1186/s41239-021-00284-9

15. Truong, A., Walters, A., Goodsitt, J., Hines, K., Bruss, C.B., Farivar, R.: Towards automated machine learning: evaluation and comparison of AutoML approaches and tools. In: 2019 IEEE 31st International Conference on Tools with Artificial Intelligence (ICTAI), pp. 1471–1479. IEEE, Portland, OR, USA (2019). https://doi.org/10.1109/ICTAI.2019.00209

16. Chauhan, K., et al.: Automated machine learning: the new wave of machine learning. In: 2020 2nd International Conference on Innovative Mechanisms for Industry Applications (ICIMIA), pp. 205–212. IEEE, Bangalore, India (2020). https://doi.org/10.1109/ICIMIA48430.2020.9074859

17. Larsen,, K.R., Becker, D.S.: Automated machine learning for business, 1st edn. Oxford University Press (2021). https://doi.org/10.1093/oso/9780190941659.001.0001

Learning Analytics for Open Learning Environments: Connection to 21st Century Skills

Sokratis Karkalas[1]([✉]), Alisa Lincke[2]([✉]), and Marianthi Grizioti[3]([✉])

[1] University of Derby, Derby, UK
s.karkalas@derby.ac.uk
[2] Linnaeus University, Växjö, Sweden
alisa.lincke@lnu.se
[3] Educational Technology Lab, National Kapodistrian University of Athens,
Athens, Greece
mgriziot@eds.uoa.gr

Abstract. This paper presents the design and development process of an innovative learning analytics tool tailored to address the challenge of generating analytics for diverse, constructionist, open learning tools. The paper outlines the preparatory phase, which informs a co-design component aimed at eliciting data requirements and establishing a shared conceptual framework among educators regarding the monitoring of 21st-century skills using open learning tools. Through this process, common understandings and agreed-upon metrics for assessing skill cultivation were identified. Subsequently, the paper delineates how these insights were translated into functional and technical specifications for the development of a learning dashboard. Finally, a report is given on the results of a preliminary evaluation, giving promising indications that the designed tool effectively addresses the identified challenges. This paper contributes to the advancement of learning analytics by offering a systematic approach to designing tools that support the cultivation of 21st-century skills in diverse learning environments.

Keywords: Learning Analytics · Requirements Elicitation · 21st-century Skills · Open Learning Environments · Exploratory Learning Environments

1 Introduction

Current international education agendas and directives underscore the imperative of digital transformation in education, advocating for a shift towards a trans-disciplinary, inclusive, and skill-oriented approach. Emphasis is placed on cultivating digital competences and emerging 21st-century skills. Contemporary educational strategies aimed at fostering these skills integrate interactive and constructionist learning media, fostering dynamic learning environments. However, this integration complicates instructional design, presenting a multifaceted

© The Author(s) 2025
T. E. Kim et al. (Eds.): MIS4TEL 2024, LNNS 1274, pp. 23–36, 2025.
https://doi.org/10.1007/978-3-031-84170-5_3

and challenging task. Within this landscape, Learning Analytics (LA) plays a central role in enhancing instructional design and optimizing teaching and learning processes. In this context a typical top-down implementation of LA would neglect the nuanced needs of such environments without early-stage involvement from educators. While existing approaches involve stakeholders in the development of LA systems, challenges persist, especially in the critical early stages. One significant challenge is the varying levels of awareness among stakeholders regarding the potential of LA, which can hinder their effective engagement in the development process. To address these challenges within constructionist environments and 21st-century skill development, this study proposes the co-design, development, and evaluation of a customisable LA dashboard. This dashboard incorporates input from teachers regarding the collection, analysis, and visualization of data generated from students' engagement with various tools. The paper aims to present a structured methodology for the design of a customisable visualization dashboard, deriving implications for conceptual models of 21st-century skills to make LA meaningful for end-users.

2   Literature Review

The integration of digital technologies into educational practices has transformed the way learning experiences are delivered and assessed [5,9]. Interactive and constructionist learning environments, which encourage active engagement, collaboration, and exploration, have become increasingly prevalent [21]. However, the complexity of these environments presents unique challenges for instructional design and assessment. Traditional methods of evaluation may not adequately capture the diverse forms of learning and engagement facilitated by these environments, necessitating the need for more evidence-based approaches such as Learning Analytics (LA).

LA can offer a holistic view of the learning process, capturing a wide range of data including student interactions with learning materials, performance on assessments, and engagement patterns [3,17]. By analyzing this data, educators can gain valuable insights into student learning behaviors, identify areas for improvement, and tailor instructional strategies to meet individual needs. This can help educators monitor student progress in real-time, providing timely interventions and support when necessary as well as summative feedback after the activities are completed. Moreover, LA enables revisiting learning designs and informed redesign adjustments to improve effectiveness in response to how designs are perceived by learners [12,16].

In the context of 21st-century skills development, which emphasizes critical thinking, creativity, collaboration, and digital literacy, LA plays a crucial role in assessing and fostering these competencies [1]. By analyzing student interactions with digital tools and resources, LA can provide valuable feedback on the acquisition and application of these skills [6,6]. Moreover, LA can inform the design of learning experiences that promote the development of 21st-century skills, facilitating a more dynamic and adaptive approach to education.

Despite its potential benefits, the successful implementation of LA requires collaboration and engagement from all stakeholders, including educators, students, administrators, and policymakers. A top-down approach to LA implementation may overlook the nuanced needs of diverse learning environments, highlighting the importance of early-stage involvement from educators in the development process. By adopting a co-design approach, which incorporates input from teachers in the design and development of LA tools, we can ensure that these tools are tailored to the specific needs and contexts of open learning environments [2, 22].

Learning Analytics offers a promising avenue for enhancing instructional design, optimizing teaching and learning processes, and fostering the development of 21st-century skills. By leveraging data analytics techniques and adopting a user-centered design approach, we can develop innovative LA tools that support the cultivation of these essential competencies in diverse learning environments.

3 Context

This project took place in the context of a large EU funded project - ExtenDT2 - focused on exploring the fusion of emerging technologies with existing constructionist learning environments, coupled with design thinking methodology, to enhance educational practices [18]. To facilitate experimentation with these technologies, we developed an innovative learning platform as a web-based learning ecosystem enabling seamless integration and interoperability of diverse learning environments. These environments vary in architecture, application programming interfaces, communication protocols, and data formats. However, within this platform, they function cohesively, offering dynamic synthesis of engaging learning activities spanning a wide array of 21st-century skills.

At the core of this platform there are interactive constructionist learning environments, fostering knowledge acquisition through exploration. These environments can be enhanced dynamically with automated support and adaptability. As learners engage with these activities, the system captures and analyzes their interactions, providing valuable insights for both learners and educators. These insights serve as formative and summative feedback for learners, aiding in self-awareness of their learning progress and refinement of learning strategies. Educators can utilize this data to assess and refine their instructional designs, fostering iterative improvement.

While this platform presents compelling learning opportunities, it also poses significant challenges, stemming from the diversity and exploratory nature of the learning process. Integrated learning tools include MaLT2 (a programming environment for creating and tinkering with 3D dynamic graphical models), SorBET (a tool for authoring classification games), ChoiCo (a tool for authoring choice-driven simulation games), and GearsBot (educational robotics).

4   Methodology

This section outlines the systematic approach employed to address the objectives given in Sect. 1. It provides a detailed account of the procedures, techniques, and tools utilized throughout the design process. The methodology adopted follows user-centered design (UCD) principles [19] aimed at uncovered needs, goals, tasks, preferences, challenges, and behaviours of end-users and stakeholders and effectively transform them into technical and functional specifications for the development of a learning analytics tool. The basic principles behind this approach involve focus on users, iterative process, multidisciplinary collaboration, early and continuous user involvement and empirical evaluation. The methodology involves the following components:

4.1   Selection of Learning Activities

UCD emphasises the importance of involving stakeholders early and continuously throughout the design process. This typically involves conducting user research, gathering feedback on prototypes, and engaging users in co-design activities to ensure that their needs and preferences are adequately addressed. This first component is related to the latter. In the context of this project we implemented a series of requirements elicitation workshops with teachers to explore what type of information they find important when students interact with exploratory learning tools with regard to 21st century skills. This was based on the Repertory Grid Technique (RGT) [20]. The workshops took place at the National and Kapodistrian University of Athens (NKUA) in June (Malt2) and November (SorBET, ChoiCo) of 2023. Further details about these workshops are given in [10]. In the first part of these workshops, participants are presented with a simplified scenario in which they assume the role of teachers utilizing a specific tool for a classroom activity. The aim is to ensure that all participants share a common understanding of the data generated by the learning tool throughout the learning process. This exercise facilitates the establishment of a uniform level of awareness regarding the semantics of the data, enabling participants to identify any gaps and introduce new, potentially valuable data elements. The facilitators are responsible to resolve possible misconceptions and make sure the same level of understanding is achieved at the end of this session. This presupposes that the facilitators are equipped with adequate level of understanding themselves before engaging with participants. This is an important factor that can influence significantly the effectiveness of this part. To address this issue we went through an elaborate process of analysing existing and well tested learning activities for each tool from old repositories. These are existing scenarios for activities based on these learning tools, that have been used in the classroom successfully and there is available documentation about misconceptions, common problems, landmarks, typical solutions and other important aspects of the learning process. The outcome of this process was the best candidate learning activities for each tool. These activities were then used to prepare facilitators for the workshops. The selected learning activities were the following: The activity for the MaLT2 tool is called "squares to cubed". It provides students with a cube net model created by a Logo procedure and asks them to modify the code so that the cube net can fold into a full cube [14]. The activity for the ChoiCo tool is called "CT-chef" and asks the students to play a simulation game about running a restaurant with healthy food and then modify it to improve its performance [7][1]. The activity for the SorBET tool is called "App-game" and asks students to play and extend a classification game about popular computer applications and their common purpose of usage [8][2].

Rationale Behind the Structure of Action Indicators: As explained in 4.1 facilitators need to be aware of the particularities of the learning tools as well as the data generated by them during activities. In this section we are presenting, as examples, two learning tools with central role in the project: ChoiCo [15] and SorBET [8]. These are web-based applications that allow the design and play of digital games by integrating a set of interconnected computational affordances. ChoiCo integrates a map-based editor, an interactive database and block-based programming, while sorbet integrates a game scene, an interactive database and block-based programming. Both applications support two modes for the users: a) the "play mode", where they can play or test a game as players with limited access to the game affordances e.g. in ChoiCo they can only see a representation of the data of a selected record from the database but not modify them and b) the design mode, where they can modify and create new game elements using the full functionalities of the integrated affordances e.g. in ChoiCo they can modify all elements of the game data in the interactive database. In the logging process we aimed to capture a) the role under which the student interacts with the tool (i.e. player or designer) and b) the different uses of the offered affordances. To achieve this, the logging messages for ChoiCo and SorBET tools are structured as follows:

- **id**: an integer variable determining the id of the relevant affordance e.g. in ChoiCo there are 3 coding editors and this id determines which of the 3 generated the event
- **type**: a string variable with the affordance that actions were performed on. It can have one of the following values: playmode (to determine that the user was interacting as a player), map_editor, database, codespace (these three values determine that the user was interacting as a designer)
- **event**: a string variable with the name of the event that is related to the element (e.g. 'addField', 'modify_point').
- **state**: any information related to this event including a) the current state of the related affordance or the game progress in case of playmode events e.g. the total number of database fields at the time of the event, b) information of the user activity that triggered the event e.g. the name of the new field that the user added and triggered the event and c) the times this event has been triggered since the start of the activity

---

[1] http://etl.ppp.uoa.gr/choico/?CTChef_Eng.
[2] http://etl.ppp.uoa.gr/sorbet/?AppGame.

– **timestamp**: the time of event trigger in Unix timestamp

For example an event from SorBET, triggered when the user, as a game designer, modifies the name of a column in the database that represents a game category could reveal the following: Through the state variable we could see that the game, at the time of the event, had X categories in total and that the student changed the name of the category 'Mamals' to 'Fish'. Regarding the learning process, this indicates that the student probably decided to use more focused categories and not so general ones like the initial game had (Mamals). This can also be considered as a landmark for the activity since it indicates an understanding of the classification model of the game and how the game categories are represented in the database.

Examples of Activities: In the following ChoiCo activity, students are asked to first play a 'half-baked game' about having a balanced diet and then improve it. The initial game has by design some intentional errors in the values of some of the choices, aiming to trigger students to discuss the game topic and correct them in the design mode. For instance, the consequences of the point 'ice cream' are not consistent with all the other similar points (Fig. 1).

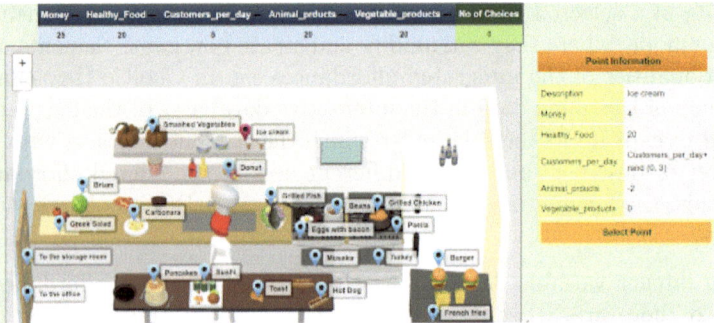

Fig. 1. A game in ChoiCo about making balanced food choices.

The teacher expects students to play the game several times, explore all the available points and then switch to edit mode and start modifying the game database. For this specific activity, the focus is on the consequence values that are represented in the database, rather than the game rules in the codespaces or the game scene. Thus the teacher indicates the landmarks shown in the first column of Table 1 and they can be correlated with the logging activity shown in the second column.

4.2   Exploration of Learning Activities

In the first part of the RGT workshops, participants are given as input a learning scenario and a full list of events associated with the respective tool. The source

Table 1. Landmarks and Events

Landmark	Event Logging Activity
Exploration of available choices as players	(The event 'point selection' has been triggered at least 20 times AND The event 'point selection' has been triggered for at least 10 different points) OR (The event 'game over' has been triggered at least 3 times with score greater than 13)
Testing and debugging	The event 'switch mode' has been triggered at least 5 times
Experimentation with game data	The event 'change_value' has been triggered at least 15 times OR The event 'add_field' has bee triggered at least 2 times
Detection and correction of inconsistencies	pointID: 14 and 22 (these are the points with the inconsistent values)

for this list was the tool documentation. The goal in this initial segment of the workshop, is to ensure uniform awareness among participants regarding the data generated by the tool throughout the learning process. The aim is to establish a common, shared understanding of the data semantics and empower participants to identify and introduce any new, potentially valuable data elements that may be missing. The scenario that is given as a point of reference is inevitably an element that influences the discussion around the data generated during the activity. This data is expected to be a proper subset of the full list of the events given and is also expected to have some activity-specific semantic nuances. It is important that the facilitators are very well prepared for that part to ensure cohesion and consistency of the semantics used in the workshop. To address this need we employed a tool called AuthELO (Fig. 2).

AuthELO [11,13] is an existing, established and well tested tool that can be used to configure user activity logging rules and rules that dictate how automated feedback can be generated based on these logs. We used AuthELO to configure the learning tools to generate all available events and then asked the facilitators to do the activities as students through AuthELO. This workshop took place in early June 2023 at NKUA with three participants. All the participants were experienced learning technologists with different backgrounds (primary school teacher, ICT and maths teacher). The participants spent 30 min for each activity - tool, including a 5–10 minute discussion / reflection.

AuthELO is designed to operate as an example-tracing tutor. It allows the user to do an activity as a learner and displays in real-time user activity indicators generated by the tool related to that activity. This allows the user to map much easier the actions performed on elements in the learning environment

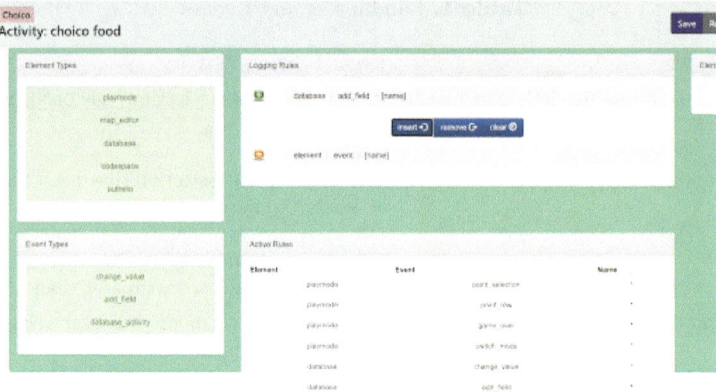

Fig. 2. Logging Rules.

to events generated by the tool and understand better their context specific nuances. The feedback we received from the facilitators for this was very positive. They reported that doing the activity themselves and seeing what is being generated in terms of events in real time streamlines the familiarisation process and provides a more exact depiction of what the actual data representation is in relation to the contextual semantics of the activities (Fig. 3).

Fig. 3. Action Indicators.

4.3   Elicitation of Data Requirements and Emerging Concepts

The outcome of the two previous components are used as input for this part. This is the series of workshops employed to uncover and analyze teachers' personal constructs, to serve as the cognitive framework through which we interpret and understand what is meaningful and useful in terms of data for the given activities. As described in [10] by utilizing this methodology, our goal was to systematically and comprehensively understand stakeholder perceptions, preferences, and constructs. This, in turn, allowed us to derive evidence-based, well-informed, and meaningful user-centered design specifications for a learning analytics tool tailored to the specific context of learning. To achieve this, we conducted a series of workshops with teachers to elicit requirements. These workshops aimed to explore what information teachers consider important when students interact with exploratory learning tools in the context of developing 21st-century skills. Through this process, we identified various concepts associated with 21st-century skills and translated them into tool-specific indicators of learner interactions. The main concepts identified include motivation, experimentation, understanding, interaction with technology, and originality and are consistent across all three tools used in the study. As illustrated in Fig. 4, each concept is linked to one or more 21st-century skills and accompanied by a list of tool-specific events that demonstrate how learners engage with the tools during learning activities. In this context, we can consider these concepts as an additional variable or dimension in the generated data, providing a categorisation perceived by educators. These categories can be mapped to specific skills and corresponding learner actions. These insights formed the foundation of a design specification, informing the development of a customisable learning analytics tool suited for open learning environments.

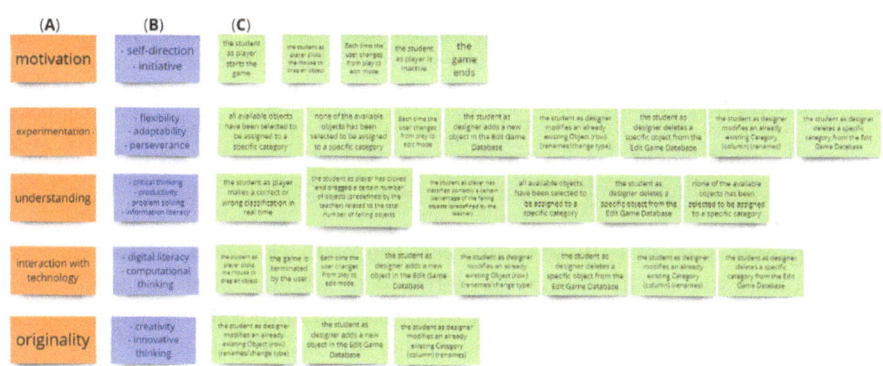

Fig. 4. Data Design for SorBET. (A) Key common concepts identified through RGT associated with (B) specific 21st-century skills and (C) tool-specific events.

4.4 Definition of Technical Specifications

The following methodology was followed to translate the outcomes of the RGT into technical specifications for the development of the learning analytics tool:

- **Understanding of RGT results**: Reviewing concepts, skills, action indicators and their interrelationships as identified in Sect. 4.3. Concepts are second-order variables in this analysis. These are the actual variables deemed worth monitoring by teachers. The values given to them are composite and should be generated dynamically as a function of the first-order action indicators specified in the RGT analysis. In this part a distinction was made between action indicators being available directly in the data and indicators that need to be generated as aggregate, derivative values from the data.
- **Mapping Action Indicators to Elements/Events**: The first-order action indicators considered in the previous step, are associated with events being generated when actions are performed on specific elements in the learning tools (e.g. move the camera or slider, execute the code, change properties of an item). Some of those indicators correspond to count-based events which quantify various learner actions, such as the number of times a student executes a procedure in MaLT2, or selects a point during the game in ChoiCo, or the number of correct/wrong classifications in a SorBET game. In this step, all of those action indicators were mapped to the actual tool-specific element/event combination.
- **Quantifying Concepts**: The concepts considered in step one are still qualitative data that need to be quantified. These are second-order variables and thus their values should depend on the respective first-order action indicators. In this part certain algorithms were considered as to how this translation should be made so that it allows easy manipulation (authorability) by teachers. The algorithm prevailed was the composite index value (score) as a weighted sum of first-order action indicators [4]. This is easy to comprehend by users and easy to represent in authoring tool user interfaces. In a composite index where multiple variables are combined with different weights, the sum of the weights should be equal to 1 or 100%. This guarantees that the composite index reflects the relative importance of each variable in the overall measurement. Each weight represents the proportion of influence that the corresponding variable has on the final composite score, in this case the concept.
- **Designing Authorability**: In this part the feedback collected from the RGT component was qualitatively considered to determine the affordances of the learning analytics tool with respect to the level of authorability. The following requirements were identified: (a) a teacher should be able to see the overall score for each concept (e.g., motivation, experimentation, etc.); (b) a teacher should be able to select a construct and configure it by enabling / disabling the respective action indicators and specifying their significance level between 0 and 100%; (c) a teacher should be able to see the elements through simple visualizations at different levels of granularity (group or classroom level).
- **Selecting Visualisations**: Simple bar charts and matrix-like visualizations were selected due to their simplicity, expressiveness and analytical power.
- **Integration with the ExtenDT2 platform**: The tool was designed to be developed as an external component and integrated with the ExtenDT2 platform in a loosely coupled manner through a REST (Representational

State Transfer) interface. The design objective here was to allow for maximum versatility, autonomy and adaptability to different deployment settings.

4.5 Development of the Prototype

The prototype was developed using the Angular 16 (TypeScript) framework[3] with the Plotly[4] and ng2-charts[5] visualisation libraries. It consists of two main components: services (for loading and pre-processing the LA data) and visualisations (for visualising the processed LA data). The business logic of the LA is implemented in the former. This involves loading and translating the data, aggregating the events, and calculating composite index scores for concepts (Fig. 5).

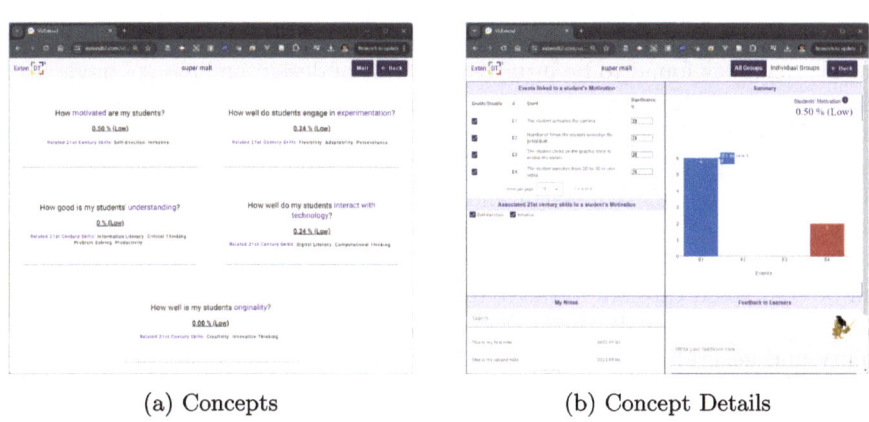

(a) Concepts   (b) Concept Details

Fig. 5. Learning Analytics Tool.

4.6 Evaluation of the Prototype

Early assessment of the prototype by stakeholders resulted in highly positive feedback. The implementation was found to closely match teachers' expectations regarding essential aspects to monitor during the learning process. Moreover, users found the interface intuitive, both in design and navigation.

---

[3] https://angular.io/.
[4] https://plotly.com/javascript/.
[5] https://valor-software.com/ng2-charts/.

5   Conclusion, Future Work

In this study, we have presented a systematic approach to the design and development of a customisable Learning Analytics (LA) tool tailored to address the challenges of generating analytics for diverse, constructionist, open learning environments. Through a co-design process involving educators, we elicited data requirements and established a shared conceptual framework for monitoring 21st-century skills using open learning tools. The data requirements derived from the design phase were effectively transformed into functional and technical specifications for the development of the LA dashboard. These specifications guided the implementation process, leading to the creation of a prototype that integrates input from teachers regarding the collection, analysis, and visualisation of data generated from students' engagement with various tools. An early evaluation of the prototype by stakeholders yielded very positive feedback. The implementation was found to be well-aligned with teachers' perceptions of what is valuable to monitor from the learning process. Additionally, the usability of the interface was found to be intuitive in terms of design and ease of navigation. This study contributes to the advancement of learning analytics by offering a structured methodology for designing tools that support the cultivation of 21st-century skills in diverse learning environments. By incorporating input from educators throughout the design and development process, we ensure that the resulting LA tool meets the specific needs and contexts of open learning environments. Moving forward, further refinement and validation of the prototype will be conducted through iterative testing and evaluation with a wider range of stakeholders. Additionally, future research efforts will focus on exploring the scalability and generalisability of the LA tool across different educational contexts and settings.

Acknowledgments. This study was co-funded by the European Union (Grant agreement: 101060231) ExtenD.T.2: Extending design thinking with emerging digital technologies, https://extendt2.eu/. Views and opinions expressed are however those of the author(s) only and do not necessarily reflect those of the European Union. Neither the European Union nor the granting authority can be held responsible for them.

References

1. Buckingham Shum, S., Crick, R.D.: Learning analytics for 21st century competencies. J. Learn. Anal. **3**(2), 6–21 (2016)
2. Chalvatza, F., Karkalas, S., Mavrikis, M.: Communicating learning analytics: Stakeholder participation and early stage requirement analysis. In: CSEDU 2019-Proceedings of the 11th International Conference on Computer Supported Education, vol. 2, pp. 339–346. SCITEPRESS–Science and Technology Publications (2019)
3. Dawson, S., Joksimovic, S., Poquet, O., Siemens, G.: Increasing the impact of learning analytics. In: Proceedings of the 9th International Conference on Learning Analytics Knowledge, pp. 446–455 (2019)

4. Diamantopoulos, A., Winklhofer, H.M.: Index construction with formative indicators: an alternative to scale development. J. Mark. Res. **38**(2), 269–277 (2001)
5. Fischer, G., Lundin, J., Lindberg, J.O.: Rethinking and reinventing learning, education and collaboration in the digital age–from creating technologies to transforming cultures. Int. J. Inform. Learn. Technol. **37**(5), 241–252 (2020)
6. Gašević, D.: Using learning analytics to measure 21st-century skills (2019)
7. Grizioti, M., Kynigos, C.: Children as players, modders, and creators of simulation games: A design for making sense of complex real-world problems: Children as players, modders and creators of simulation games. In: Proceedings of the 20th Annual ACM Interaction Design and Children Conference, pp. 363–374 (2021)
8. Grizioti, M., Kynigos, C.: Integrating computational thinking and data science: the case of modding classification games. Inform. Educ. **23**(1), 101–124 (2024)
9. Haleem, A., Javaid, M., Qadri, M.A., Suman, R.: Understanding the role of digital technologies in education: A review. Sustain. Oper. Comput. **3**, 275–285 (2022)
10. Karkalas, S., Chalvatza, F., Mavrikis, M., Nikolaou, M.S.: Using the Repertory Grid Technique in a Co-design Process for Learning Analytics: Conceptualisation and Connection to 21st Century Skills. In: Proceedings of the 14th International Conference on Methodologies and Intelligent Systems for Technology Enhanced Learning, MIS4TEL 2024. , Salamanca, Spain (June 2024), in press
11. Karkalas, S., Mavrikis, M.: Feedback authoring for exploratory learning objects: Authelo. In: CSEDU 2016-Proceedings of the 8th International Conference on Computer Supported Education. vol. 1, pp. 144–153. Science and Technology Publications, Lda (2016)
12. Karkalas, S., Mavrikis, M., Labs, O.: Towards analytics for educational interactive e-books: the case of the reflective designer analytics platform (rdap). In: Proceedings of the Sixth International Conference on Learning Analytics and Knowledge. pp. 143–147 (2016)
13. Karkalas, S., Mavrikis, M., Xenos, M., Kynigos, C.: Feedback authoring for exploratory activities: the case of a logo-based 3d microworld. In: Computers Supported Education: 8th International Conference, CSEDU 2016, Rome, Italy, April 21-23, 2016, Revised Selected Papers 8, pp. 259–278. Springer (2017)
14. Kynigos, C., Grizioti, M.: Programming approaches to computational thinking: Integrating turtle geometry, dynamic manipulation and 3d space. Inform. Educ. **17**(2), 321–340 (2018)
15. Kynigos, C., Grizioti, M.: Modifying games with choico: Integrated affordances and engineered bugs for computational thinking. Br. J. Edu. Technol. **51**(6), 2252–2267 (2020)
16. Mavrikis, M., Karkalas, S.: Reflective analytics for interactive e-books. IxD&A **33**, 33–53 (2017)
17. Michos, K., Lang, C., Hernández-Leo, D., Price-Dennis, D.: Involving teachers in learning analytics design: Lessons learned from two case studies. In: Proceedings of the Tenth International Conference on Learning Analytics and Knowledge, pp. 94–99 (2020)
18. Milrad, M., et al.: Combining design thinking with emerging technologies in k-12 education. In: International Conference in Methodologies and intelligent Systems for Techhnology Enhanced Learning, pp. 15–27. Springer (2023)
19. Norman, D.A., Draper, S.W.: User centered system design; new perspectives on human-computer interaction. L. Erlbaum Associates Inc. (1986)
20. Rozenszajn, R., Kavod, G.Z., Machluf, Y.: What do they really think? the repertory grid technique as an educational research tool for revealing tacit cognitive structures. Int. J. Sci. Educ. **43**(6), 906–927 (2021)

21. Shah, R.K.: Effective constructivist teaching learning in the classroom. Online Submission **7**(4), 1–13 (2019)
22. Vezzoli, Y., Mavrikis, M., Vasalou, A.: Inspiration cards workshops with primary teachers in the early co-design stages of learning analytics. In: Proceedings of the Tenth International Conference on Learning Analytics and Knowledge, pp. 73–82 (2020)

The e-DIPLOMA Platform:
A Cloud-Based Solution for Educational Groupwork in Gamified Environments

Viktória Burkus[1]([✉]), Andrea Castelli[2]([✉]), Ádám Sike[1], László Szécsi[1]([✉]),
and Gergely Tomcsányi[1]([✉])

[1] Department of Control Engineering and Information Technology, Faculty of
Electrical Engineering and Informatics, Budapest University of Technology and
Economics, Műegyetem rkp. 3., 1111 Budapest, Hungary
`{burkus,sike,szecsi,tomcsanyi}@iit.bme.hu`
[2] Brainstorm, Avenue de la Albufera, 321., Madrid 28031, Spain
`acastelli@brainstorm3d.com`

Abstract. Distributing gamified educational content, including media
and executables, and conducting multi-user groupwork activities are
challenging because of technological, logistical, ethical, and financial concerns. We address these questions by proposing a cloud-based solution
that allows safe and controlled management of educational activities
without the need for an on-premises server array. While technology to
manage student and educator data, authentication, and grading exist
in the forms of LMSs, extending these to online multiplayer games and
virtual reality environments is lacking. We introduce a framework integrated with an LMS capable of managing such games, which also includes
fine-grained handling of user consent regarding access and use of assets.
We also explain how the setup ensures safe and secure data management,
especially when recording sensitive data in educational experiments. We
detail the design of the web interface that allows teachers and students
to set up groups that work together in separate virtual environments,
backed by dynamically allocated cloud servers, with automated acquisition of all prerequisite software and assets. An evaluation session with
15 students is analyzed to verify whether the system can meet the above
goals in the time constraints of a class activity.

Keywords: Educational Technology · Groupwork · Gamification ·
Cloud · LTI · LMS

1 Introduction

Multiplayer online games have been long heralded as the future of education [1].
More recently, large-scale studies concentrated on individual multiplayer games,
e.g. the Radix Endeavor [2]. In some contexts like teaching programming, already
digital development tools can be retooled for co-creation and learning [4]. However, it remains an immense hurdle to establish infrastructure for such a game.

© The Author(s) 2025
T. E. Kim et al. (Eds.): MIS4TEL 2024, LNNS 1274, pp. 37–48, 2025.
https://doi.org/10.1007/978-3-031-84170-5_4

Existing popular game environments, e.g. Minecraft, can be repurposed to some extent for some applications, but those are not integrated with any Learning Management System. This means identities, rights, privacy, data protection for students, and privileges for teachers cannot be easily managed, and student evaluation cannot happen in-game. Therefore, we developed a platform that manages the required online infrastructure in the cloud, is integrated with the LMS, and provides an online interface for organizing students into groups.

The e-DIPLOMA Platform is a framework for cloud-based management of interactive educational computer applications, related assets and education materials, as well as a means for distributing and executing those applications on user and cloud devices with lobby-based matchmaking for group activities, and monitoring activities for research and profiling purposes. It is compatible with existing systems used in education through LTI [8], but also brings high-end gaming, virtual and augmented reality technologies into the framework.

The platform was designed to host the prototype courses of the e-DIPLOMA project [9], and facilitate their teaching in pilot courses, in accordance with ethics and values laid out in previous research[5]. Therefore, it is essential that it supports the evaluation of activities performed. These prototype courses address the subjects of block programming, social entrepreneurship, and VR skills. All activities are realized by applications, mostly implemented over Unity or Unreal Engine, but shared use of Edison [10] AR environment, coupled with video conferencing, or web-based games are also possible. Some activities are single user, but our focus in this paper is on multiplayer or supervised games that require a network connection. These include a shared VR city simulation, co-operative business planning using canvas templates, and a competitive product and marketing management board game with simultaneous actions and computer-assisted management of the hidden market variables.

To that end, it provides mechanisms for the storage and management of acquired data, including server activity logs and client-side measurements. It manages the granting and revoking of consent, and stores data in a safe and secure way in accordance with the consent given. Users have well-defined roles, with permission to perform certain actions tied to those roles.

The cost of cloud services is controllable by the instructors and managers operating the system. Installation on user devices is designed to be simple, and updates are automatic. The platform enables the use of game engines, integrating applications that allow for real-time graphics and physical simulation, VR and AR visualization, and AI, which can be considered disruptive technologies in education [3].

2   Functionality

2.1   Underlying Technologies and Services

The platform uses the Moodle learning management system [7]. The e-DIPLOMA service is implemented in AWS, hosted as a set of web pages on the cloud, communicating with other AWS services for authentication (Cognito), data storage (S3), serverless computations (API Gateway, Lambda), game server operation (GameLift) and Elastic Computing System (EC2).

Moodle is a certified LTI 1.3 [8] platform. The e-DIPLOMA service integrates with Moodle using eLTI [6]. Thus, e-DIPLOMA users keep their identities and roles defined by their institutional Moodle accounts, and need only to log in to Moodle once. The *e-DIPLOMA Frontend* consists of the web pages providing the user interface for e-DIPLOMA users. The *e-DIPLOMA Backend* is a set of serverless operations (AWS Lambda) accessible through websocket and http endpoints (AWS API Gateway). They store data in the AWS DynamoDB database, and rely on S3 buckets for file storage. The *e-DIPLOMA Launcher* is a desktop application that be installed on a Windows computer. *AWS EC2* is a service that allows the use of computer instances in the cloud. Applications, and Brainstorm's Edison [10] in particular, can be deployed to EC2 instances directly. *AWS GameLift* is a managed way to run game servers on EC2 instances. The e-DIPLOMA applications that are based on a client-server game engine model run their game servers on GameLift.

2.2 Concept

Activities are items visible on Moodle course pages, added by a Teacher (acting in the role of e-DIPLOMA Course Administrator). e-DIPLOMA activities are a specific type of activity. Clicking such an activity opens an e-DIPLOMA Frontend webpage. e-DIPLOMA activities execute some *learning objects*. Learning objects are organized into *courses*, e.g. the *prototypes* created for the e-DIPLOMA project. A learning object is a unit of learning material, based on a *version* of an *application*, accompanied with associated *assets*. Assets are collections of files managed by the e-DIPLOMA platform. Assets can be part of learning objects, but they can also be private files of users, the result of work or co-creation performed by users, activity logs or measurement data. Permissions regarding these and all other resources are managed using *consent forms*. Through a consent form, the owner of a resource can permit a well-defined set of users to list, view, execute, contribute to, edit, or delete them. A consent form lists the person(s) responsible for the appropriate handling of the data with their contacts, purposes and risk associated with the data use, and the users receiving the authorization through the consent form. Data owners can grant their consent to the terms of the consent form pertaining to any of their data resources individually, or revoke that consent.

2.3 Platform Overview

Figure 1 shows the overview of the e-DIPLOMA Platform. Black boxes symbolize pre-existing infrastructure, key interfacing standards are in grey, and orange boxes are high-level components of the platform.

Users log in to *Moodle* using the method preferred in their institutes. Often, this means using a global service (e.g., log in with Google, or login with Facebook), or an institute-wide solution. Once users are logged in to Moodle, they

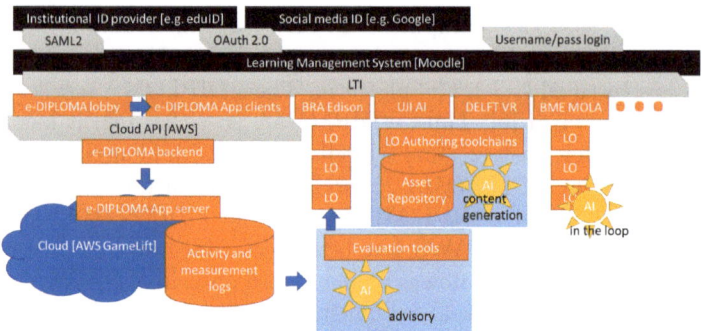

Fig. 1. The e-DIPLOMA Platform with its environment and application features.

can access courses according to their roles. *External Tool* activities can be added to course pages by content editors or teachers and launched by students. The *e-DIPLOMA Lobby* is such a Tool. Moodle and the e-DIPLOMA Lobby are interfacing using the LTI protocol.

The e-DIPLOMA Lobby is a web service for launching *Learning Objects* (LOs), including organizing events or sessions where multiple users take part. While some LOs may be implemented as web services themselves, the need for using high-end technologies like AR, VR, and gamification would typically require specialized desktop applications built on extensive middleware, and game engines in particular. Practically, a single application should be used to implement many LOs. Typical game AI like pathfinding may be accompanied by disruptive AI like chatbots to provide a responsive and immersive experience. It is foreseen that e-DIPLOMA partners each implement one or a few applications, and several LOs within each application, as part of the prototype course development process. Adopters in the exploitation phase of the project would replicate this development process to integrate their own curricula.

For multiplayer scenarios (which includes not only group work but also real-time teacher-student interaction), application instances running on user devices must be connected to create a synchronized game world. This is provided by the application *server* component. Depending on the type of the LO, multiple simultaneous instances of the game world need to be simulated, accommodating interacting groups of students and teachers. These instances may be run on-premise servers, but typically they need to be hosted in the AWS cloud. Launching these dedicated server computers in the cloud, managing the authentication of users joining the sessions, linking client applications to the server application, and monitoring resource usage is the task of the e-DIPLOMA Platform Backend, which is an AWS service. Finally, the server applications may record user activity relevant to academic or scientific evaluation. They can store this data in the cloud through the backend.

Figure 2 gives an overview of the e-DIPLOMA Platform components as implemented in the actual software, elaborating on platform functionality. The *Man-*

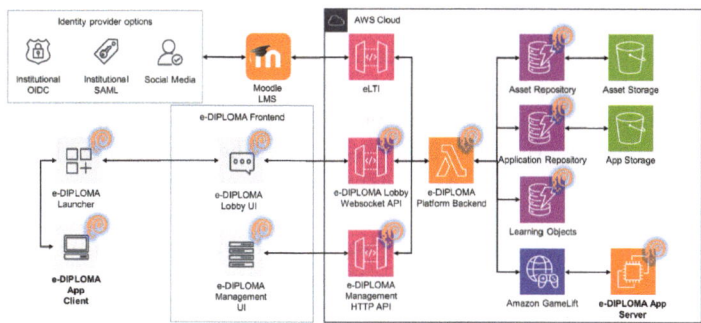

Fig. 2. The e-DIPLOMA Platform core as implemented.

agement UI and *API* are web pages with underlying infrastructure to create and manage e-DIPLOMA resources, including applications, courses, learning objects, consent forms, and assets. The *e-DIPLOMA Launcher* is an essential element that is installed on the client computers, managing the download and setup of application clients to these devices, and also launching application client instances and connecting them to the servers in the cloud. In Fig. 2, the entire chain facilitating the connections between the app clients and app servers can be followed.

A *Moodle* server is maintained by BME for the purposes of the project, including administering the pilots. Educational institutes that adopt e-DIPLOMA can use their own LMS to manage their users and perform conventional course management activities. Any authentication method supported by Moodle, preferred by or previously implemented by the institute, can be used. This includes OIDC and SAML federated authentication. The course prototypes of e-DIPLOMA will be created as Moodle courses, the pilots will manage student evaluation through Moodle.

The *Lobby* is the e-DIPLOMA Platform interface that most users, and students in particular, are going to encounter most often. Activities on the Moodle course page launch this interactive web page, where students and teachers can see users logged into e-DIPLOMA through clicking the same activity in the LMS. They can communicate via chat, organize groups if necessary, and launch an application presenting the Learning Object associated with the activity. The service performs matchmaking, gathering the users willing to join the same interactive virtual environment. Lobby usage and functionality is explained in detail in Sect. 4.2.

The Lobby must start desktop applications on client computers. This is only possible for applications installed and registered in the operating system. Therefore, the Lobby is accompanied by a *Launcher* component. The Launcher is responsible for keeping the installed e-DIPLOMA applications up-to-date and starting them when required by the lobby. Only the Launcher has to be registered with the operating system for browser access. Interaction with the Launcher is described as part of the Lobby experience in Sect. 4.2.

A part of the backend is responsible for managing the lobby connections of several students and teachers, facilitating communication between them, handling their actions like joining or leaving virtual rooms, and, most importantly, to launch game servers in the cloud via GameLift. User clients can connect to these servers. This functionality is accessed from the browser running the Lobby page. Actions by other students or teachers are shown immediately on other user's computers.

Every message sent from the browser to the backend has to be accompanied by proof of identity of the caller. Otherwise, anyone could claim access to user resources, posing a severe security risk. In case of the e-DIPLOMA, this proof is provided by Moodle through eLTI.

Applications are developed by e-DIPLOMA partners while creating the prototype courses. Later adopters may also develop applications. Most applications are expected to work in a client-server model. The server can be launched by the backend in the cloud. The server can record activity data and store it in the repository through the backend. Clients gather measurement data about the user. They can store this data into the repository through the backend.

The *asset repository* is a multi-purpose data storage facility. Besides storing and organizing media assets useful for the creation of interactive educational material, it also serves as personal data storage space for students and educators. Measurement and activity log data required for analysis are also stored as assets. Access to assets is governed by consent of their owners. All assets are accessible from the Asset Management Page (see Sect. 4.4), which is part of the frontend.

A *learning object* combines an application version from the Application repository with assets from the Asset repository.

3   User Interface

A role is a set of capabilities or privileges assigned to a user. The e-DIPLOMA Platform relies on Moodle to manage user roles. Main roles that play a part in activity execution are *student* (symbol:), supervisor (), and course manager (). This means roles can be assigned on a per-course basis. Through LTI, the cloud backend can check whether a user has a role required for an action. The front end user interface also reflects the actions allowed to the user, the most prominent example being the difference between the student and supervisor interfaces detailed later in Sects. 4.2 and 4.3. A supervisor may join students in virtual environments, monitor student activity, and control the use of cloud services. The course manager is an educator responsible for populating a Moodle course with content, including e-DIPLOMA activities.

The User Interface consists of web pages. The management pages are dedicated to the creation and editing of assets and asset groups, courses and learning objects, applications and versions. All of them can have consent forms attached, detailing what privileges other users are granted over the resources. Consent forms themselves can be edited in the Consent Forms Management Page. The Lobby page provides the interactive interface to distribute applications, organize group activities and launch educational sessions.

4   Asset Management

Every user of the platform can act as an asset owner. They can upload files (3D models, textures, images, videos, presentations, etc.) to the platform, typically accompanied by a description and a thumbnail image - creating a *media asset*. Recordings made about the actions performed, i.e. activity logs, are also assets. Similarly, measurements made are stored as assets. Every user is the owner of the assets they created.

Users can also create asset groups, and add media assets or other asset groups to them—producing arbitrary groupings of assets.

Owners of assets can enroll their assets under consent forms, thereby granting their permission to the kind of use specified, to the users specified, under the responsibility of the people specified. Through these granted permissions, assets other than the user's own can show up in listings, and can be used in appropriate ways.

Private assets are intended for either development use, or as a personal storage space for users, especially instructors and students. Learning objects may allow the use of these resources, selected and accessed by the participants when the educational activity is performed, not when the learning object is authored. A presentation delivered in a virtual environment or a 3D model created by a student can be such an asset.

4.1   Consent Management

Not only assets, but also applications, courses, learning objects need a control mechanism on access rights. The system that the e-DIPLOMA platform utilises is based on the concept that data owners need to give their informed consent on the use of their data. This is accomplished through consent forms. A consent form lists the person(s) responsible for the appropriate handling of the data with their contacts, purposes and risk associated with the data use, and the users receiving the authorization through the consent form. A user can be granted the following levels of access, each including the privileges of the preceding levels:

list the name and owner of the resource can appear in listings shown to the user

view the detailed properties of the resource can be shown to the user

execute the resource can be downloaded and uploaded and executed (if executable) or processed by an application

contribute if the resource encapsulates other resources, the user can add new elements to it

edit the user can change any properties

delete the user can remove the resource from the repository

Data owners can grant their consent to the terms of the consent form pertaining to any of their data or resources individually, or revoke that consent.

4.2   Activities as a Student

Students can access the e-DIPLOMA platform if there has been an e-DIPLOMA activity added to the course page by a course supervisor.

When the student opens (clicks on) the activity, they are redirected to the e-DIPLOMA user interface, opening the e-DIPLOMA lobby in a new tab of the browser. Users can interact with other users in the lobby via chat, and have prerequisite software and assets downloaded on their device in advance, to save time during actual class. The lobby immediately starts the e-DIPLOMA Launcher, which then downloads any missing applications and assets. If the Launcher has not been installed, the user will be prompted to download and install it.

Once the supervisor has set up virtual rooms and enabled the use of cloud servers, users can join rooms, and, upon supervisor approval, join VR, desktop, or web-based single or multiplayer sessions (Fig. 3).

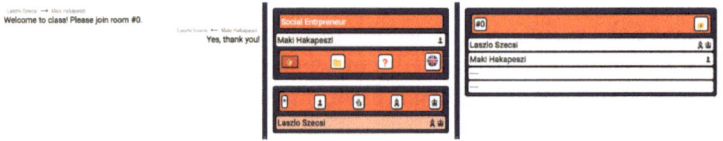

Fig. 3. Detail of the lobby populated by online users and virtual rooms. Note that students can select other users by role, can send individual or group messages, and join rooms.

The *Profile panel* shows the title of the learning object associated with the Lobby page, the card of the user logged in through the browser, and buttons with actions and options that pertain to the user. A user card always displays the full name of the user on the left side, and the icons indicating their roles on the right side. The profile panel may contain buttons with the following icons:

- : Download Launcher.
- : Bring assets to room. This button allows the user to select one of their asset groups. The assets referenced by this group will be both copied to the game server and downloaded for the game client. Thus, students can display or work with their own projects or files in the virtual environments.
- : Join session. This button only appears if the user has joined a virtual room, a supervisor has approved that room, initiated the session, and the application server has been started in the cloud. Then, by pressing this button, the user can request the Launcher to launch the client with the appropriate parameters to join the server in the cloud. Then the interactive multiplayer activity can start.

The *Participants panel* appears below the Profile panel, with the list of users logged in to the lobby. User cards can be dragged and dropped to join or leave rooms, or invite others to rooms. Users can be selected individually, by role, or by room.

The *Chat panel* shows sent or incoming messages. Messages typed in the input field are sent to selected users. The chat can be used to organize groupings and populate rooms.

The *Rooms panel* contains a number of rooms with a pre-determined number of slots, as created by the supervisor (see Sect. 4.3). Rooms may be in the following states, indicated by the icon in the upper right corner:

- : Open. Users can join the room as long as there are free slots. Users can also leave the room.
- : Locked. This indicates that the supervisor has approved the room setup, and the activity can start. Users cannot join or leave the room.
- : Waiting. The session has been initiated, and the backend is busy starting the server. Users cannot join or leave the room.
- : Active. There is a live session with a server running. Users can join the session, or have already joined the session. Users cannot join or leave the room.

Once the user has joined a room, and a supervisor locked the room and started the session, students can launch their app clients by clicking the Join session button in the Profile panel (see Sect. 4.2). The e-DIPLOMA Launcher proceeds to download prerequisites, including assets brought along (see Sect. 4.2) by users in the room. It is advisable to open the activities before the class, to download the prerequisites in advance, and not delay joining the session with last-minute downloads. From this point on, the e-DIPLOMA App is responsible for interaction with the user.

4.3   Activities as a Supervisor

A supervisor is a teacher who is responsible for presenting a learning object to students in a course, and supervising their activities in the e-DIPLOMA Lobby and in e-DIPLOMA Apps. It is the supervisor's responsibility to start the e-DIPLOMA Lobby and ensure the allocation of sufficient cloud resources before the activity starts. Using cloud resources for e-DIPLOMA App servers incurs costs. Therefore, their use should be initiated right before they are needed, and relinquished as soon as they are no longer in use (Fig. 4).

All the options available to students are also available to the supervisor, but they can also move students around the rooms by dragging their user cards. Supervisors can create and delete rooms, adjust the number of users in a room, lock rooms and start them. Most importantly, before rooms could be started, appropriate cloud capacity has to be prepared. This capacity is measured in the number of concurrent sessions, and every room started takes up such a session. Once this capacity has been prepared, the costs associated with maintaining the cloud resources apply. These costs can depend on momentary factors, and also whether cloud or on-premise servers are used. After the activity, the supervisor

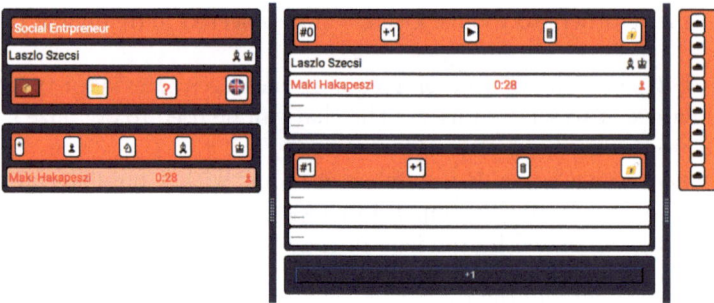

Fig. 4. Supervisor view detail of a Lobby. Note the buttons available for managing the number and size of rooms, opening, locking, starting, and deleting them.

also has to relinquish this capacity. As starting a server can take some time, a timer indicates when will be possible to start a room using that cloud server. The supervisor should allocate adequate cloud capacity right before, or at the beginning of the class. All capacity is automatically relinquished at the end time of the activity as specified in the LMS. It is most economical, however, for the supervisor to relinquish all cloud capacity as soon as it is not needed.

4.4   Other Resources

The *Learning Object Management* page displays learning objects organized by courses. They can be edited, or a new learning object can be added. Every learning object must have an e-DIPLOMA App Build specified, and possibly a client asset group and a server asset group.

A Build is a certain version of the App. Learning Objects are always authored based on a specific version. A build may consist of a client and a server executable. Core media assets may be included, but learning object assets and user files are not part of the Build—they are made accessible on the machines after the Build has been deployed to them.

5   Results and Discussion

We conducted an experiment with 14 students (7 present in a computer room, 7 online, none of them familiar with the platform) and a teacher to evaluate lobby functionality. In particular, how fast it is possible to set up the cloud service, install prerequisites to user computers, form student groups, and launch virtual environments with those student groups. Our experiment did not concern the activities and measurements after that point. Figure 5 gives a timeline of when the milestones were reached (Fig. 5).

Some participants were delayed by their virus scanner raising objections to the Launcher, and others were using Firefox browser, while only Chrome was supported. In the presence of an audio communication channel (in person or via

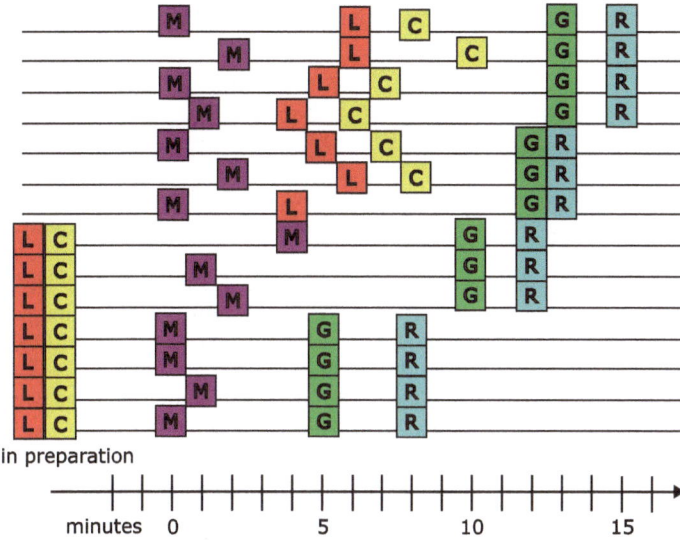

Fig. 5. Time of milestones reached, in minutes, by participant. L - Launcher installed, C - client installed, M - logged in to Moodle, G - group formed, R - virtual room launched

Microsoft Teams), the lobby chat functionality was only utilized when requested by the supervisor. We can conclude that it took at most 8 min to form groups and start the activity for unexperienced users with a pre-class login. Some students had to wait for the teacher to approve their rooms, before it could be started, so the bottleneck in our setup was the attention of the teacher. With installation of the prerequisites during class, the process was completed in 15 min.

We conclude that our system is able to provide dynamically allocated cloud infrastructure for gamified group activities. In scenarios with a higher number of students, and few teachers, the possibility to start a virtual room should not be reserved for the teacher, as it puts too much load on her, and student have to wait. Therefore, we are going to experiment with automatic approval of room setups, based on pre-configured rules chosen by the teacher (e.g. minimum and maximum number of participants, group composition by grades or other characteristics). Further research will focus on evaluating other functionalities, in particular grading, asset management, and measurements during education sessions.

Acknowledgments. Research supported by the e-DIPLOMA, project number 101061424, funded by the European Union. Views and opinions expressed are, however, those of the authors only and do not necessarily reflect those of the European Union or the European Research Executive Agency (REA). Neither the European Union nor the granting authority can be held responsible for them.

Disclosure of Interests. The authors have no competing interests to declare that are relevant to the content of this article.

References

1. Paraskeva, F., Mysirlaki, S., Papagianni, A.: Multiplayer online games as educational tools: Facing new challenges in learning. Comput. Educ. **54**(2), 498–505 (2010)
2. Ruiperez-Valiente, J.A., Gaydos, M., Rosenheck, L., Kim, Y.J., Klopfer, E.: Patterns of engagement in an educational massively multiplayer online game: A multidimensional view. IEEE Trans. Learn. Technol. **13**(4), 648–661 (2020)
3. Boucher, P., Bentzen, N., Laţici, T., Madiega, T., Schmertzing, L., Szczepanski, M.: Disruption by Technology: Impacts on Politics, Economics and Society, European Parliamentary Research Service, 2020. PE652.079 (2020). https://www.europarl. europa.eu/stoa/en/document/EPRS_IDA(2020)652079
4. Staubitz, T., Klement, H., Renz, J., Teusner, R., Meinel, C.: Towards practical programming exercises and automated assessment in Massive Open Online Courses. In: 2015 IEEE International Conference on Teaching, Assessment, and Learning for Engineering (TALE), (pp. 23-30) (2015). IEEE
5. Väljataga, T., et al.: Exploring value and ethical dimensions of disruptive technologies for learning and teaching. In International Conference in Methodologies and Intelligent Systems for Technology Enhanced Learning, (pp. 107–116) (2023). Cham: Springer Nature Switzerland
6. Enable LTI. https://aws-samples.github.io/enable-lti/. Accessed 12 April 2024
7. Documentation MoodleDocs. https://docs.moodle.org/403/en/Main_page. Accessed 29 Feb 2024
8. Learning Tools Interoperability(LTI) Advantage Implementation Guide 1.3 1EdTech Final Release | IMS Global Learning Consortium. https://www.imsglobal. org/spec/lti/v1p3/impl/. Accessed 29 Feb 2024
9. e-DIPLOMA Project. https://e-diplomaproject.eu/. Accessed 12 April 2024
10. Brainstorm Edison. https://www.brainstorm3d.com/products/edison/. Accessed 12 April 2024

Teaching and Learning with Design Thinking and Emerging Digital Technologies in K-12: Impact on Students and Teaching Recommendations

Christothea Herodotou[1]([✉]), Sagun Shrestha[1], Feiran Zhang[2], Christina Gkreka[3], Carina Girvan[4], Sofia Papavlasopoulou[2], Marianthi Grizioti[3], Eileen Scanlon[1], Chronis Kynigos[3], and Marcelo Milrad[5]

[1] The Open University, Walton Hall, Milton Keynes MK7 6AA, UK
{christothea.herodotou,sagun.shrestha,eileen.scanlon}@open.ac.uk
[2] Norwegian University of Science and Technology, Høgskoleringen 1, 7034 Trondheim, Norway
{feiran.zhang,spapav}@ntnu.no
[3] National and Kapodistrian University of Athens, 157 72 Athens, Greece
{xristgreka,mgriziot,kynigos}@eds.uoa.gr
[4] The University of Dublin, College Green, Dublin 2, Ireland
girvanc@tcd.i
[5] Linnaeus University, Hus D, 2269D Växjö, Sweden
marcelo.milrad@lnu.se

Abstract. This paper presents insights from five case studies with 173 students aged 11–14 years old across three countries in Europe, exploring the integration of design thinking (DT) and emerging digital technologies in K-12 education. The study aimed to understand the impact of DT and digital technologies on the development of students' 21st-century skills and capture the challenges students and teachers face. The focus is to give evidence-based recommendations for teaching using the DT approach. Despite the positive student reports on developing communication, collaboration and programming skills, other skills related to research and reflection, and understanding of DT were rather less developed. The main challenges teachers faced were related to time constraints, difficulties in using the technologies, lack of clear instructions, and limited understanding of the DT process. Seven practical recommendations on how to effectively teach and learn DT with digital technologies are presented. These are accompanied by the Exten. (D.T.)^2 Digital Design Thinking Model, an innovative framework designed to guide the use of DT with specific digital technologies across K-12 education.

Keywords: Design thinking · digital technologies · 21st century skills · K-12

1   Introduction

Design thinking (DT) integrates creativity and problem-solving within the curriculum to foster students' development of 21st century skills such as critical thinking, collaboration, empathy, and argumentation. Due to a focus on the iterative prototyping of solutions,

T. E. Kim et al. (Eds.): MIS4TEL 2024, LNNS 1274, pp. 49–60, 2025.
https://doi.org/10.1007/978-3-031-84170-5_5

DT has been perceived as a means for developing a resilient or "fail early and often" mindset (Marks & Chase 2019). While DT has been recognised for its great education potential and has been increasingly used in K-12, there is still limited empirical evidence supporting its effectiveness when used in the classroom (Li & Zhan 2022). Emerging evidence points to the development of collaboration, communication, metacognition and critical thinking. Critical thinking was associated with empathy and reflection emerged from defining a problem (Rusmann & Ejsing-Duun 2022).

DT faces challenges when used in the classroom related to its effective integration into existing curricula, the need to develop teachers' skills and attitudes towards DT processes, and helping students become creative and more collaborative (Andersen & Pitkänen 2019). A range of digital technologies have been used in DT, yet challenges remain and relate to teachers' management of multiple digital and non-digital technologies, balancing different modes of teaching and navigating a complex design thinking process (Hjorth et al. 2016).

In this paper, we report on findings from five pilot case studies across three countries that used DT with digital technologies in K-12 education (11–14 years old). We address the following Research Questions (RQs): **RQ1:** What is the impact of teaching using DT and digital technologies on the development of students' 21st century skills? **RQ2:** What challenges have teachers and students reported during teaching using DT and digital technologies? **RQ3:** What evidence-based recommendations for the effective implementation of DT activities emerge from RQ1 and RQ2?

This study is part of the Horizon Europe and Innovate UK funded project: Extending Design Thinking with Emerging Digital Technologies – Exten. (D.T.) [2] (https://extendt2.eu) that aims to bring methodological innovation in K-12 education by applying, testing and scaling up a DT implementation and a suite of emerging digital technologies (Milrad et al. 2023). It builds on the premise that emerging digital technologies enable effective application of DT in the classroom and bring added value to existing practices. In the next section, we present literature on DT and learning.

2 Background Literature

DT has been applied in K-12 education in solving engineering problems (Mentzer et al. 2015), engage students in programming tasks (Avcu & Er, 2020) and teach students physics concepts (Simeon et al. 2022). DT can play a crucial role in K-12 education, for example, a recent study (Lin et al. 2024) employed DT to teach IT courses to junior middle school students and found that DT helped enhance students' creativity in their digital works and improved their awareness of using IT tools to solve everyday challenges. Another study (Cutumisu et al. 2020) taught middle school students DT strategies such as critical feedback-seeking and revising.

Despite promising learning outcomes, DT can be challenging due to students needing assistance with using tools such as 3D modelling (Weibert et al. 2014), lacking active interaction with digital and physical materials limiting their ideas, needing guidance about how to analyse and externalise ideas and justify decisions, taking things for granted (Weibert et al. 2014). Similar challenges were reported by teachers, including skills on how to use digital technologies and teach DT (Smith et al. 2016), how to think more creatively and innovatively (Razali et al. 2022), how to map assessment activities to learning objectives, especially given the non-linear nature of DT in which assessment frequently takes on informal forms (Veldhuis et al. 2022), and how to accommodate the extended time needed to implement a DT project (Andersen & Pitkänen 2019).

Various tools have been designed for DT such as tools for authoring DT-based learning activities (Bekker et al. 2019) and scaffolding DT processes (Gennari et al. 2022). However, technology has not been used to cover all stages of DT; there is a lack of dedicated tools to support stages such as empathy and capture students' learning progress in real time supporting distributed or online collaboration. The potential of emerging technologies to support DT in K-12 classroom has neither been sufficiently leveraged nor robustly evaluated. In the coming sections, we present our approach to addressing some of these gaps.

3   The Exten. (D.T.)^2 Tools Used in DT

We used a set of free digital technologies to support DT implementation at schools. These tools have been developed and are managed by project partners, thus enabling us to modify and use them as and when required. They were seen to support different stages of DT: the "empathise" stage of DT is enabled by **nQuire for students** while rapid prototyping and iteration are enabled by online games: **ChoiCo, SorBET, MaLT2** and **GearsBot.**

ChoiCo (https://extendt2.com/widgets/choico/) enables students to play, modify or create choice-driven simulation games online. The player makes choices with conflicting consequences in a map-based setting. The winner is the player who survives the longest. The game can help students develop decision-making, empathy, argumentation, and systems thinking. It can be used to explore contemporary socio-scientific issues such as climate change, balanced diet. An example of a game activity is the "Covid-19 risks" asking students to make choices of daily activities that will not lead them to catching Covid-19 (Grizioti et al. 2021).

SorBET (https://extendt2.com/widgets/sorbetExt/) enables students to design, modify, and play Tetris-like classification games. Players using the mouse, or their body movements must quickly classify falling objects into the categories they belong to. Sor-BET allows design and modification of game elements such as density, speed, rules, object and category definitions, with block-based programming and database affordances. It can help students to develop 21st century skills such as classification skills, critical thinking, and computational thinking.

MaLT2 (https://extendt2.com/widgets/malt/) allows students to create, animate and print 3D figural models with text-based programming and dynamic manipulation. The models can vary from simple cubes to complex DNA models, jewels and fractal trees. Ideas and concepts from mathematics, engineering, art and computer science are seamlessly combined in a creative and collaborative process of experimentation, tinkering and self-expression. An example activity is the creation and then printing of 3D jewels by students as a means to understand the importance of the degradation of material.

GearsBot (https://extendt2.com/widgets/gears/) is an open-source 3D robotics simulator with two programming and one simulation tabs. One of the programming tabs has a simple form of programming, namely block-coding that users can drag-and-drop blocks. The other tab requires that the users write their code in Python. An example activity is that students use a virtual "world" about "fire rescue" and try to improve how the robot is constructed and controlled to complete a rescue mission.

nQuire for students (https://learn.nquire.org.uk) is an online platform that enables students to design, pilot and manage research studies. Data can be collected from other students and teachers who are registered with the platform. A classroom management system allows teachers to create classes and student accounts and navigate within and between different schools. nQuire for students allows students to collect data in the form of text, numbers, images, and sensors such as light and sound data.

4   Methodology

Methods of Data Collection: In this paper, we report on five case studies in Country 1 (x2), Country 2 (x2), and Country 3 (x1). Each case study describes a school intervention occurring in the classroom, as part of the school day or after-school activities with students aged 11–14 years old. Within each case study, two groups of consented students were chosen to act as "focal groups". Ethical approvals were gained for all five case studies. Pre-intervention and post-intervention questionnaires (RQ1) were used as the primary medium to capture self-reported impact on learning and 21st century skills. Data were also collected through observations (video, screen recordings, audio and/or semi-structured written), small group interviews, and teacher interviews (RQ2 and RQ3). Interview data were transcribed and translated to English where applicable.

Methods of Data Analysis: The phases of analysis were as follows: a) constant comparative analysis to identify issues most pertinent to participants (interviews and observational data. b) Critical incidents were then used to identify events that were significant in the action and to explore them in depth (observations and interviews). c) Survey data were analysed using descriptive statistics to provide additional evidence and content to (a) and (b).

Contextual Information Per Case Study: Case study 1: Country 1 Case study 1 was conducted in an international private school with 58 students aged 11–14, most of whom had prior knowledge of programming with Scratch. The study was carried out in three classes. In each class, one researcher acted as the facilitator, while one teacher supported all actions as needed. The DT topic was: "How to improve cybersecurity and online awareness through a choice-and-consequence-based game", and technologies used: nQuire for students and ChoiCo. The DT sessions lasted 6 h (2 days x 3 h). Data were collected from 14 student interviews, 58 pre-intervention questionnaires, 56 post-intervention questionnaires and 3 teacher interviews. Case study 2: Country 1 Case study 2 was conducted in a public school with 15 students 15–16 years old, who had previous knowledge of block-based and Python programming. The study was carried out in a single classroom with one teacher overseeing and two researchers facilitating the sessions. Two one-hour-long sessions spanned 3 weeks (for a total of 2 h). The topic of the DT project was: "How to improve fire safety awareness through an educational virtual robot?" and the technology used was GearsBot. Data were collected from 4 student interviews, 15 pre-intervention questionnaires, 8 post-intervention questionnaires and 1 teacher interview.

Case study 3: Country 2 The study was conducted in a public junior high school. Thirty (N = 30) students from the after-school club Mathematics and Programming, aged 14–15 participated in the study for 8 h (4 sessions x 2 h per week. A teacher and a researcher jointly facilitated these sessions. The topic of the DT project was to design a 3D digital model of a vertical garden watering system. Technologies used were MalT2 and nQuire for students. Data were collected from 23 pre-intervention questionnaires, 26 post-intervention questions, 2 student interviews, 8 audio-visual recordings, and 1 teacher interview. Case study 4: Country 2 This study was conducted in a state-funded secondary school. Three classes of 67 students, aged 13–15 years old participated in the study for 8 h (4 sessions × 2 h). The role of the teacher was to provide support as needed. The topic of the DT project was: "Playing with Environmental Issues" and students used ChoiCo and nQuire for students. Data included 67 pre-intervention questionnaires, 59 post-intervention questionnaires, 3 student interviews, 8 audiovisual recordings, and one teacher interview.

Case study 5: Country 3 This study was conducted in a public secondary school. One third of students were eligible for free school meals. A class with 23 students, aged 11 to 13 participated. Four sessions of 45 min were conducted (3,5 h). The teacher facilitated the sessions. The topic of the DT project was saving electricity at school and technologies used were Choico and nQuire for students. Data were collected from 18 pre-intervention questionnaires, 7 post-intervention questionnaires, 2 focus group interviews, 8 audio-visual recordings, and screen recordings of students' interactions and discussions.

5 Findings

To address RQ1, we analysed questionnaire data: a) Pre-intervention questionnaires: 173 responses from students aged 11–14. 52% of them were male. 85% of them reported using desktop computers/laptops regularly and 92% smartphones. 25% used technology to create digital art and videos. The remaining used it to search the web, watch videos and play games. b) Post-intervention questionnaires: 152 responses of which 56.3% were male. Students agreed that they had participated in the various elements of a DT project. They agreed that they had 'helped make decisions about what their group would create (M = 4.33), had 'tested and improved the solution' (M = 4.23). Less than half agreed that they had 'found out about the needs of other people' (42.6%), with most (43.3%) choosing 'neither agree nor disagree' for this statement. The statement with the most negative responses - 'I was involved in presenting the solution to other people' (18.8% disagree) may reflect the fact that some DT activities were shortened due to time constraints. Most students agreed (M = 4.35) that they had worked well as part of a team. Students' responses (N = 135) to what skills they developed after taking part in a DT project revealed that cooperation was the most mentioned skill, followed by programming, teamwork, thinking and communication. Students stated confidence to use these skills in future projects. The least developed skills were research, reflection, patience, and design thinking.

To address RQ2, we analysed interview data and observation notes taken during the DT sessions. Case study 1 and 2 (Country 1): The challenges students reported were related: a) technology's limited functionalities: challenges were observed while students were coding with ChoiCo due to limited functionalities: "the coding part is a bit hard because it has a limited number of variables you could choose from and a limited amount of stuff you could make with it." b) limited time: A student commented: "I feel like we had some time, but there was a lot, there was not enough time to make a good game." Likewise, another group of students mentioned: "We didn't have a lot of time to think about a question before we had to move on to the next [session].", c) ideation: As explained: "Challenges faced… for me, it was about ideas. Like I put my best in trying to think of ideas of what to do for phishing and scamming." and d) task management: while tasks were easy to do, these were many making the process of keeping track of them rather overwhelming: "Making the code was a bit confusing, but not like [design thinking and coming up with choices]. It was easy, but just a lot of [things to do]. So, you have to keep track of everything."

The challenges teachers reported were related to: a) limited time: one teacher said, "We could have spent some more time showing them how the program works, or maybe there could have been a few examples of what you were sort of looking for. […] But, like here, we kind of gave them not that much time to really get into it." and b) the DT process: understanding what to do at each phase of the DT process was seen as another challenge: "Obviously, there were technical challenges, but that comes up all the time. I think the biggest challenge they had was trying to navigate how to solve the problems. Some other challenges they have[…], is knowing what it was that they were being asked to do in each stage".

Case study 3 and 4 (Country 2): The challenges students faced during the DT process were related to: a) limited time: students expressed a sense of dissatisfaction from the inability to fulfil initial objectives: "If time were more properly managed and we finished the project, it would be one of the most fun! It's a shame to put in so much effort and not see it completed!", b) understanding how to use technology: prior experience proved to be essential regarding the time students needed to finalise their designs. When students were not familiar with the technology tools, this has a negative impact on the learning process. A student mentioned: "It's not particularly difficult for someone who hasn't used it before, but it takes time and requires guidance." c) collaboration: as a student explained: "We had difficulties in communication, because we were a bigger team (i.e., three students instead of two) and we had disagreements about the values of the game that led to tensions between us." Students resolved these disagreements either by exploring a new solution or compromising: "We used to be more opinionated in the beginning. But now, if we had opposing opinions on a game element, we would come up with a third solution." d) the DT methodology: difficulties reported related to the open-ended nature of DT and feedback activities. Regarding the former, a student explained," The most challenging part of the process was the actual design phase, we had to consider various parameters before starting to design in Malt2. Factors such as the placement of the model in the garden, its shape, and the choice of materials all needed to be carefully combined to achieve a successful construction outcome". Though challenging, students acknowledged that this aspect made the process interesting: "The most difficult part was when we had to decide on the numbers of the game. It wasn't difficult. It was the cleverest part; the most challenging". During the feedback phase, the main issue that emerged was that students' comments on their peers' designs were not specific enough to enable them to make changes to their creations. A student explained, "we were told that our game was boring. So, what should we change? This comment didn't help us.", while another student added to this: "We respect their opinion and maybe our game is too easy for them, so they get bored when they play it. But when we asked them to explain to us how we could improve it, they said again that it was a boring game. So, we didn't understand what we could change to make it more interesting."

Accordingly, teachers faced challenges related to a) limited time, b) students' prior experience with technology and c) assessment of students' learning outcomes. Teachers stressed that the implementation of the DT methodology in a formal school context would require significant time, making it difficult to integrate it into the existing school programme unless the process is streamlined and well prepared. A teacher stated: "the time was not enough. I think that the duration of the activity plan was defined correctly. However, this time was enough for the game and the activity on the game, for the peer review, for the reconstruction of the game and in general for the actual stages of the design thinking. [...] it was not enough for other stages.

Regarding students' prior experience with technology, another teacher said: "With MalT2 students had worked with before while nQuire had not. This had the effect of delaying the first phase quite a bit as it is a tool with many and open functionalities. They have to see it beforehand. I had to make a questionnaire for them to work on and direct them on what kind of questions they would make. The original design was changed as at first they were going to make it entirely by themselves". Finally, another important issue teachers mentioned was the difficulty to assess the learning outcomes of their students. When asked what skills he thinks the students have developed during the project, a teacher replied "I am not able to know in such a short and limited time precisely what kind of skills or competencies were developed and if students have developed any of them. Furthermore, a single project may not be enough to be able to extract confirmed results.

Case study 5 (Country 3): The challenges students faced were mainly related to a) students' engagement in DT sessions and b) the use of project technologies. The disengagement of some students was a major concern. As explained the instructions on how to engage in the DT sessions were not clear. In the first focal group, students highlighted that their classmates were disengaged from the activity most of the time. Both groups reported that there was a lack of instructions, which could potentially guide them as to how to use the project technologies. As explained: "the teacher could just [give] more instructions for the game so that we can understand better." Also, students expressed the need for individual support from the teacher to get explanations for tasks: "I found it hard trying to fill out the questionnaire [on nQuire for students], fill those questions for people to answer about my game [...]I can't really think about any questions. Overall, there was a need for clear instructions on how to use technology.

Regarding the use of project technologies, students in both focus groups discussed the fact that games created in SorBET and ChoiCo had to be downloaded on local computers and others had to access them through these computers (instead of e.g., these being saved online and accessed via a URL). Students expressed very strongly that it was not feasible for their classmates to play the games as they could not upload the games online and then share it with others. As explained: "The game wouldn't download for me. So I had to just keep remaking it." In another example using ChoiCo, students had difficulties using the pins on a picture as these would immediately disappear. For nQuire for students, they found the interface not friendly enough as they had to scroll all the way down to select the response type after typing a question.

An interview was not conducted with the teacher, therefore challenges reported below emerged from the observation of lessons. There was limited time available for the teacher to implement the DT sessions. The teacher was new to both DT activities and project technologies while at the same time she was under pressure to complete the DT sessions within 4 to 5 sessions. This meant that she had to rush the implementation, skipping parts of it.

6 Discussion

In this paper, we reported on five case studies that took place across three countries, with 173 students in total, aged 11–14 years old. In these case studies, we used a set of technologies to enable delivery of DT in K-12 in an effort to understand how digital technologies could add value and facilitate effective implementation of DT in formal educational contexts. Students reported that they contributed to making decisions in their groups and that they tested and improved solutions. Yet, most of them did not find out about the needs of others or had an opportunity to present their work to others, revealing the stages of DT where emphasis was placed by teachers during the implementation. While early findings evidence impact on students' development of 21st century skills (as reported by students), several challenges were reported by both students and teachers while engaged with the DT process. We used these insights to produce a set of practical and evidence-informed recommendations for teachers to consider when designing, delivering and managing DT projects in formal education.

Regarding RQ1 and aligning with existing studies (Rusmann & Ejsing-Duun 2022), students reported development of social skills in the form of teamwork, cooperation and communication, as well as thinking, programming skills and increased confidence. In contrast, students were less likely to report development of research and reflection skills, and understanding of design thinking, necessitating the need to design DT activities that enable development of specific skills. For example, developing reflective skills would enable students to participate and find creative ways of learning and help them assume another person's perspective (empathy) (Rusmann & Ejsing-Duun 2022). The provision of feedback to other students, reported as a challenge in this study, would be an effective means to develop critical thinking skills. Critical feedback was shown to develop in DT studies with middle school students (Cutumisu et al. 2020).

Regarding RQ2 and RQ3, a major challenge noted across case studies was the availability of time to carry out DT projects in formal settings. Although DT activities were interesting for students, teachers had to rush between DT stages likely affecting students' engagement and understanding of activities. This finding with existing studies (Andersen & Pitkänen 2019) points to the significance of considering additional time when implementing DT projects in classrooms. Therefore, we recommend: **Recommendation 1:** Break down the DT process to separate sessions and factor in additional time to accommodate students' needs when interacting with the process. **Recommendation 2:** Identify which activities can be completed as homework by students to reduce time needed in the classroom.

Another challenge reported by teachers is the connection of DT activities with specific learning outcomes and ways of assessing these, a challenge also reported in the literature (Veldhuis et al. 2022). The use of a lesson plan as a preparatory tool for designing a DT project would enable teachers to reflect on how specific activities may relate to specific learning objectives and allow time to design formative and summative assessments to evaluate these. We recommend: **Recommendation 3:** Map activities to specific learning outcomes and identify assessment approaches for evaluating each.

Teachers and students' familiarity with using technologies and their functionalities emerged as another challenge which affected students' engagement in DT projects. Students required detailed guidance on how to use technologies (Weibert et al. 2014) and time to become familiar with tools functionalities. We recommend: **Recommendation 4:** Organise separate lesson plans in which students are asked to experiment with technology tools, as preparation for delivering a DT project. Effective engagement with the DT processes was also an obstacle for students. This related to a number of aspects including a) the implementation of certain stages of DT, such as the ideation stage which requires creativity and production of new ideas (Weibert et al. 2014), 2) decision making in groups aligning with existing studies that noted difficulties in reaching a consensus during for example the prototype development (Smith et al. 2016), 3) providing actionable feedback to others' creations that can help improve their work (Smith et al. 2016), and 4) managing multiple tasks within a single session. We recommend: **Recommendation 5:** Use a DT visual (model) to pinpoint to students the DT stage they work on, relevant tasks and timelines per activity. **Recommendation 6:** Give students opportunities to understand and practice decision making in groups, as preparation for delivering a DT project. **Recommendation 7:** Give students opportunities to provide feedback to the work of others, as preparation for delivering a DT project.

In response to the need of students to better understand each stage of the design thinking process (Recommendations 5 and 1) as well as the limited impact recorded on students' learning of design thinking processes, we produced a design thinking model for use in K-12 education, coined as the **Exten. (D.T.)^2 Digital Design Thinking Model** (see Fig. 1). While it builds on existing DT models, our model's unique contribution is that it details what students should be doing at each DT phase (through prompting questions) and which digital tool/s they should be using. It can be used to support teachers in designing activity or lesson plans for DT as well as a navigational tool to pinpoint to students the DT phase they are addressing. The starting point of the model is a challenge or a problem students decide to engage with while the end point is a range of outcomes including students' physical or digital creations, the development of new skills and knowledge and behaviour changes. The model features five phases as follows: empathise and understand, define and ideate, rapid prototyping and iteration, sharing and feedback, and respond and deliver. It has been integrated with an activity plan we produced to help teachers plan DT projects (https://extendt2.eu/dt-activity-plan-template/).

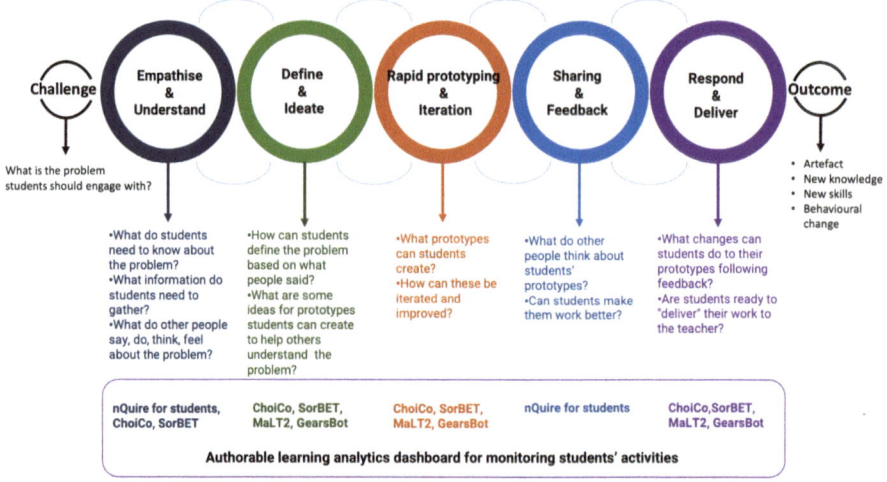

Fig. 1. The Exten. (D.T.)^2 Digital Design Thinking Model

7   Conclusions

In this study, both teachers and students reported facing challenges, notably in terms of time availability for implementing DT sessions and the readiness of the digital tools. Addressing the issue of time constraints, our approach involves clarifying to participating teachers the importance of allocating sufficient time to each stage of the DT process for successful implementation. Additionally, we aim to assist teachers in distributing learning activities across various sessions by collaboratively identifying tasks suitable for individual completion at home and those best suited for group work in the classroom. By doing so, we aim to minimize the amount of school time required to complete a DT project without compromising any essential aspects of the process.

Insights collected from this pilot implementation informed the redesign and further development of digital tools mentioned before with the aim to be user friendly and function smoothly. In addition, the tools have been extended to accommodate features including gesture-based interaction (SorBET), geolocation (Choico), 3D printing capabilities (MalT2), and scaffolding to provide feedback to research studies (nQuire for students). We aim to test, iterate and improve the new versions of the digital tools and Exten. (D.T.)^2 Digital Design Thinking Model (see Fig. 1) in the second and third years of the project, through collecting data from students and teachers while implementing DT projects. In addition, we aim to address two limitations of this study: a) analyse data considering for the sample's heterogeneity e.g., different ages, backgrounds, school type, locations, enabling comparisons between e.g., different types of schools, and an understanding of the conditions under which the proposed intervention may work (or not) well, and b) collect data to identify how and why the functionality of each technology triggers the development of specific 21st century skills.

Acknowledgments. This project has received funding from Horizon Europe and Innovate UK under grant number 10106023. This paper reflects the authors' view, and the Horizon Europe and Innovate UK are not responsible for any use that may be made.

Disclosure of Interests. The authors have no competing interests to declare that are relevant to the content of this article.

References

Andersen, H.V., Pitkänen, K.: Empowering educators by developing professional practice in digital fabrication and design thinking. Int. J. Child-Comput. Interact. **21**, 1–16 (2019). https://doi.org/10.1016/j.ijcci.2019.03.001

Avcu, Y.E., Er, K.O.: Design thinking applications in teaching programming to gifted students. J. Educ. Technol. Online Learn. **3**(1), 1–30 (2020)

Bekker, T., Taconis, R., Bakker, S., d'Anjou, B.: Developing an online authoring tool to support teachers in designing 21st century design based education in primary school. In: B. M. McLaren, R. Reilly, S. Zvacek, & J. Uhomoibhi (eds.), Computer Supported Education, pp. 142–165. Springer (2019). https://doi.org/10.1007/978-3-030-21151-6_8

Cutumisu, M., Schwartz, D.L., Lou, N.M.: The relation between academic achievement and the spontaneous use of design-thinking strategies. Comput. Educ. **149**, 103806 (2020). https://doi.org/10.1016/j.compedu.2020.103806

Gennari, R., Matera, M., Melonio, A., Rizvi, M., Roumelioti, E.: The evolution of a toolkit for smart-thing design with children through action research. Int. J. Child-Comput. Interact. **31**, 100359 (2022). https://doi.org/10.1016/j.ijcci.2021.100359

Grizioti, M., Oliveira, W., Garneli, V.: Covid-19 Survivor: design and evaluation of a game to improve students' experience during social isolation. In: International Conference on Games and Learning Alliance, pp. 283–288. Springer International Publishing, Cham (2021)

Hjorth, M., Smith, R., Loi, D., Iversen, O., Christensen, K.: Educating the reflective educator: design processes and digital fabrication for the classroom. In: Proceedings of the 6th Annual Conference on Creativity and Fabrication in Education (2016). https://doi.org/10.1145/3003397.3003401

Li, T., Zhan, Z.: A systematic review on design thinking integrated learning in K-12 education. Appl. Sci. (2022). https://doi.org/10.3390/app12168077

Lin, L., Dong, Y., Chen, X., Shadiev, R., Ma, Y., Zhang, H.: Exploring the impact of design thinking in information technology education: An empirical investigation. Think. Skills Creat. **51**, 101450 (2024)

Marks, J., Chase, C.: Impact of a prototyping intervention on middle school students' iterative practices and reactions to failure. J. Eng. Educ. **108**, 547–573 (2019). https://doi.org/10.1002/jee.20294

Mentzer, N., Becker, K., Sutton, M.: Engineering design thinking: high school students' performance and knowledge. J. Eng. Educ. **104**(4), 417–432 (2015). https://doi.org/10.1002/jee.20105x

Milrad, M., et al.: Combining Design Thinking with Emerging Technologies in K-12 Education. In: Kubincová, Z., Caruso, F., Kim, Te., Ivanova, M., Lancia, L., Pellegrino, M.A. (eds.) Methodologies and Intelligent Systems for Technology Enhanced Learning, Workshops - 13th International Conference. MIS4TEL 2023. LNCS, vol. 769. Springer, Cham (2023). https://doi.org/10.1007/978-3-031-42134-1_2

Razali, N.H., Ali, N.N.N., Safiyuddin, S.K., Khalid, F.: Design thinking approaches in education and their challenges: a systematic literature review. Creat. Educ. **13**(07), 2289–2299 (2022). https://doi.org/10.4236/ce.2022.137145

Rusmann, A., Ejsing-Duun, S.: When design thinking goes to school: A literature review of design competences for the K-12 level. Int. J. Technol. Des. Educ. **32**(4), 2063–2091 (2022)

Simeon, M.I., Samsudin, M.A., Yakob, N.: Effect of design thinking approach on students' achievement in some selected physics concepts in the context of STEM learning. Int. J. Technol. Des. Educ. **32**(1), 185–212 (2022). https://doi.org/10.1007/s10798-020-09601-1

Smith, R.C., Iversen, O.S., Veerasawmy, R.: Impediments to digital fabrication in education: a study of teachers' role in digital fabrication. Int. J. Dig. Literacy Dig. Competen. **7**(1), 33–49 (2016)

Veldhuis, A., Xiao, D., Bekker, T., Markopoulos, P.: Model-based support for authoring Design-based Learning and Maker Education materials in elementary education. In: 6th FabLearn Europe/MakeEd Conference 2022, pp. 1–9 (2022). https://doi.org/10.1145/3535227.3535230

Weibert, A., Marshall, A., Aal, K., Schubert, K., Rode, J.: Sewing interest in E-textiles: analyzing making from a gendered perspective. In: Proceedings of the 2014 Conference on Designing Interactive Systems, pp. 15–24 (2014). https://doi.org/10.1145/2598510.2600886

Application of Image Recognition in Nautical Simulator Training Assessment

Sahil Bhagat[1]([⊠]) [iD] and Ziaul Haque Munim[2]([⊠]) [iD]

[1] Department of Technology and Safety, University of Tromsø (UiT),
The Arctic University of Norway, Tromsø, Norway
`sahil.bhagat@uit.no`
[2] Faculty of Technology, Natural and Maritime Sciences, University of South-Eastern Norway,
Horten, Norway
`ziaul.h.munim@usn.no`

Abstract. Traditional methods for assessing student performance in maritime simulator training usually rely on instructor observations or post-manoeuvre debriefing. This study explores the potential of applying image recognition technology to assess student performance during critical manoeuvres such as *Williamson turn* in a nautical simulator training. By utilizing image recognition through convolutional neural network algorithm, a system is proposed that can analyze key manoeuvre aspects using visual data from the simulator, thus aiding students in assessing their performance. Analysing these visual cues with an image recognition algorithm could potentially serve as a component of a learning analytics dashboard (LAD) for maritime simulator training.

Keywords: man overboard · Williamson turn · nautical training · nautical simulator · image recognition

1 Introduction

"Man overboard" (MOB) is an exclamation used onboard a vessel when a crew member or passengers falls off the ship and into the surrounding water. It signifies a critical emergency where the person who fell overboard is in danger of drowning or hypothermia. The most severe consequence of man falling overboard is the risk of death. Almost every type of marine vessel has recorded MOB occurrence [1–3]. Cruise ship accidents involving passengers going overboard are tragically common. Statistics show that around 22 people fall overboard each year, with a high rate (79%) either going missing or not surviving [4]. The biggest hurdle for maritime search and rescue (SAR) is swiftly and accurately locating objects in the vast ocean [5]. A crew member's immediate awareness of a "man overboard" (MOB) event significantly improves the chances of a successful rescue. To minimize causalities from MOB incidents and streamline SAR operations at sea, the International Civil Aviation Organization (ICAO) and the International Maritime Organization (IMO) have jointly developed and published a manual titled "International Aviation and Maritime Search and Rescue (IAMSAR)" [6]. This manual provides guidance for both in need of assistance during maritime emergencies and those who might be able to offer aid [7].

© The Author(s) 2025
T. E. Kim et al. (Eds.): MIS4TEL 2024, LNNS 1274, pp. 61–70, 2025.
https://doi.org/10.1007/978-3-031-84170-5_6

The high seas may seem vast and open, but for a ship's captain navigating busy harbours and narrow waterways, precision manoeuvring is paramount. One such crucial manoeuvre is *Williamson turn* as shown in Fig. 1, a technique used to efficiently turn a ship within a limited space. Mastering this turn requires a combination of theoretical knowledge and practical skills. Maritime simulator training plays a vital role in honing these abilities in a safe and controlled environment. However, traditional assessment methods in simulators often rely on instructor observations, lacking a continuous and objective measure of performance [8]. Objective measures in a simulator training can provide standardized and unbiased evaluations of a trainee's skills [9, 10].

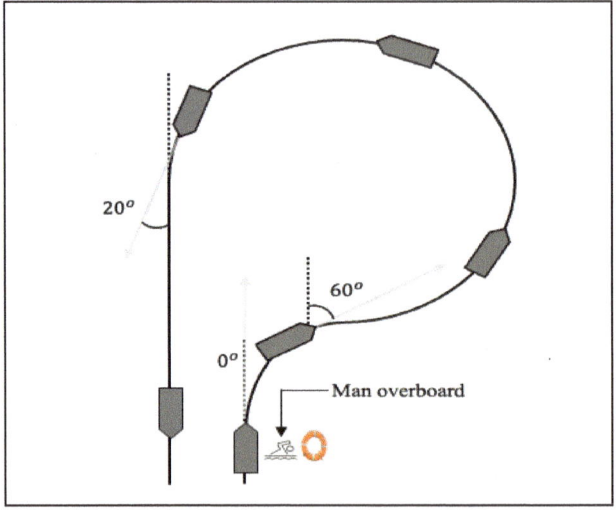

Fig. 1. Williamson turn as per IAMSAR [7]

Learning analytics dashboards (LAD) offer an objective and data-driven approach to assess performance in maritime simulator training [11]. Maritime simulator exercise generates various datasets such as simulator logs, visual vessel tracking data from Electronic Chart Display and Information System (ECDIS), Automatic Radar Plotting Aid (ARPA) and audio recordings from radio communications. Investigations into performance assessment in maritime simulator training have employed multiple datasets from simulator training and eye-tracking [12–14]. Researchers have recommended exploring the use of diverse data sets and various data analysis techniques within LADs to enhance the learning experience for students [11, 15]. LAD can incorporate visual representations of image-based assessments alongside traditional metrics, providing a holistic view of student progress and areas for improvement. By analysing visual cues captured during *Williamson turn* exercise using image recognition algorithms, we can unlock a potential to evaluate student performance. This approach has the potential to move beyond subjective instructor evaluations by providing a data-driven analysis of student success in completing the turn. This shift from a traditional instructor evaluation to data-driven insights paves the way for a personalized and effective training experience for maritime students. Therefore, the objective of this study is to explore applications of image recognition algorithms for evaluating student performance in simulator training.

2 Exercise Scenario

The *Williamson turn* is a manoeuvre used in maritime navigation, to retrace the vessel's wake as closely as possible, making it ideal for searching in low visibility [6, 7]. Specifically designed for use on rescue teams (rescue units, vessels, and aircraft), IAMSAR's volume III (mobile facilities) offers guidelines for search and rescue operations including those impacting the rescue team itself [7]. IAMSAR also recommends the Anderson and Scharnov turns as alternative recovery manoeuvers. The Anderson turn, ideal for ships with responsive handling and ample power, is the quickest method to return to the person overboard. The Scharnov turn, like the *Williamson turn*, guides the ship back into its wake, but it is not suitable for immediate action scenarios. As per IAMSAR, *Williamson turn* manoeuvre involves a rudder hard over to the side of the casualty on receiving the information of a man overboard. When the vessel has deviated 60 ° from the original course, the rudder must be shifted to the hard over on the opposite side. Finally, on reaching the vessel heading to 20 ° short of the opposite course, the rudder needs to be set at midship/'zero' position. This set of commands will help the ship to return to the opposite course from the point of commencement of manoeuvre. While this manoeuvre excels at maintaining the course in the opposite direction and can be effective when moving the ship away from the incident, its execution can be slow [16].

The dependability of ship-handling simulations is of immense importance, so we used desktop-based simulators [16]. The Kongsberg desktop-based navigation and manoeuvring simulator was used in this study. This simulator utilizes a complex mathematical model to deliver realistic responses for the trainee in their simulated environment. Additionally, it is Det Norske Veritas (DNV) certified, signifying its approval for use in specialized training courses for professional crew.

The simulation exercise constituted of a cargo ship running full ahead at a course of 050° in the open sea at 16 knots when a MOB alarm is raised, and a MOB buoy is dispatched for rescue. The weather is stable with good visibility and no current. First, the participants were familiarized with *Williamson turn* using printed instructions from IAMSAR's volume III. Thereafter, they were requested to perform a *Williamson turn* and return to close proximity to the MOB buoy as shown in Fig. 2. In this study, a total of eight students from the nautical bachelor program and one simulator instructor took part in the simulation training exercise. Exercises performed by nautical students were used as a testing set. A simulator training instructor was requested to perform the exercise scenario to generate a training sample. Also, the i-MASTER dataset was accessed and used as a training set [17]. The exercise lasted for approximately 4–6 min. Images from ECDIS of manoeuvres performed by participants were collected as samples. The samples were labelled by the instructor using the proximity distance from MOB buoy and the resemblance to *Williamson turn*.

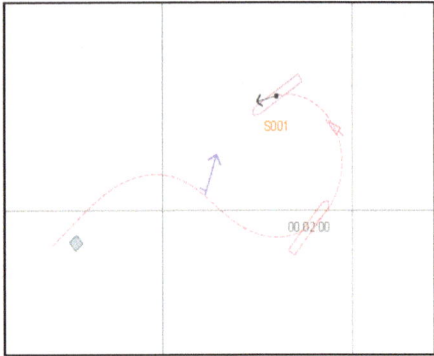

Fig. 2. Williamson turn on desktop simulator with an orange smoked MOB buoy. *Left*: Student Display screen, *Right*: ECDIS display.

3   Methodology

3.1   Overview of Convolutional Neural Network (CNN)

The instructor performed two sets of simulation exercises which are used as a training set for modelling. Images of all simulations performed by instructors along with samples collected from the i-MASTER dataset were labelled as 'high proximity' and 'low proximity' by instructors, depending on the probability of rescuing the personnel considering the final position of ship. Labelled images of *Williamson turn* performed were then used as a training data set for image recognition.

Image recognition is a computer vision technique that allows machines to interpret and categorize the content of images and visual inputs. It involves the use of algorithms and machine learning models to detect and identify objects, features, or patterns within an image. This technology is widely used in various applications, such as facial recognition systems, autonomous vehicles, medical image analysis, and photo tagging on social media platforms [18–21]. Object recognition models are used in Autonomous Sea Surface Vessels (ASV) to recognize nearby vessels, using a radar guided camera, infrared images using thermal cameras [22]. Traditionally, object recognition relied on manually designed features [19]. With the rise of Convolutional Neural Networks (CNN) and their performance in the ImageNet Challenge has solidified neural networks as the preferred solutions for general objection recognition tasks [23].

CNN is used in this study to extract useful features. CNNs leverage convolutional layers to capture spatial relationships between pixels in an image. These layers use small grids (filters or kernels) that scan the image, looking for basic building blocks like edges, lines, and shapes [24]. Pooling layers are added to down-sample the feature maps by applying a pooling operation to reduce their dimensionality and computational complexity. CNNs learn features in stages [24]. The initial layers identify fundamental building blocks like edges and shapes. Subsequent layers take these basic features and progressively combine them into more complex ones, ultimately recognizing object parts or even whole objects within the image. The final stage involves fully connected layers. These layers, like those in traditional neural networks, take the processed data from the convolutional and pooling stages and transform it into actionable insights. This allows the network to classify the image content or pinpoint the location of specific objects.

3.2   Model Architecture

In this study, *TensorFlow* library is used for developing neural network for classifying images. Figure 3 shows an overview of CNN model used for analyzing images. 4 layers of convolution layers are utilized with a final flattened, dense layer of Deep Neural Network (DNN). Rectified Linear Unit (ReLU) is used as activation function and SoftMax function is used as the ultimate layer for classification of images. During model compilation, loss is computed using 'categorical_crossentropy' function and optimizer 'rmsprop' is used to reduce losses and improve the learning rate for better accuracy.

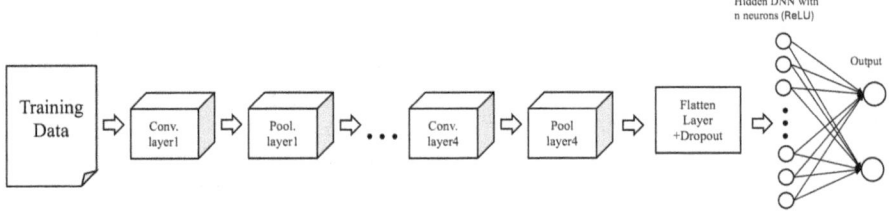

Fig. 3. Overall structure of CNN+DNN model (Convolution, Pooling, and DNN), where convolution and DNN layers needed training.

3.3   Data Pre-processing

The training and validation dataset of images has been downscaled to a uniform resolution of 300x300 pixels to standardize the data and enhance computation speed. *Keras* library is exploited for data pre-processing. *ImageDataGenerator* is a tool from the library of *Keras* which is employed for data augmentation. It pre-processes and transforms the data on the fly during training, thereby generating variations of existing images.

3.4   Model Training

The model was trained on 56 images with 28 each in two categories (i.e., Low proximity and High proximity) extracted from the instructor's execution of the simulation scenario. Model was initially trained for 100 iterations to evaluate the performance on loss and accuracy metrics. The final model was trained on 75 iterations with 64 neurons as a penultimate layer of DNN. To improve modelling time, training was carried out using mini batches. We used a computer with an Apple M1 processor and 16GB of memory to run CNN model.

3.5 Model Testing

The test set was composed of images collected from simulation exercises performed by 8 bachelor students. The same exercise was used by all participants and manoeuvre images were recorded using a computer and camera to evaluate model classification prowess. The test images were also reduced to 300-pixel size on the fly during testing.

4 Results

The training data sets consisted of 64 images, in which 56 images were used as training set and 8 images were used as validation set. Each dataset was a mixture of mobile captured photos and simulation-generated images to present a diverse quality of dataset. The final model was trained with a different size of DNN layers for 75 iterations. From the Table 1, it is evident that a DNN model of 64 layers performed faster as compared to 128 and 256 layers. This is due to lower number of training parameters in a less dense network. However, the accuracy (Fig. 5) and loss (Fig. 4) indicates limited improvement with a denser layer. A small improvement in loss and accuracy for training set was shown for DNN with 256 layers but the accuracy achieved by all models for validation set is same. Figure 4 indicates no overfitting or underfitting as both training and validation loss exhibited similar learning pattern. A sharp spike in loss is expected due to use of basic optimizer and dividing dataset in batches for faster training.

Table 1. Loss and accuracy for different DNN for Image classification

DNN Layers	Parameters	Training Set		Validation Set		Training Time
		Loss	Accuracy	Loss	Accuracy	
64 × 2	66,1762	0.448	78.57%	1.1	87.5%	1 m 41 s
128 × 2	1,063,362	0.4472	78.57%	0.549	87.5%	1 m 45 s
256 × 2	1,866,562	0.3874	82.14%	0.7142	87.5%	1 m 52 s

Figure 6 presents 8 images that were tested with the trained model to verify image classification. The images under "High Proximity" shows exercises where the participants performed *Williamson turn* and the ship's final location was in closer proximity to the MOB buoy. On the contrary, "Low Proximity" indicates exercises where the ship ended up in a position from where the rescue probability has relatively dropped. Furthermore, the observed pattern in low proximity category did not resemble the typical *Williamson turn.* Out of eight images, six were classified correctly by the trained model and two images under "Low Proximity" were misclassified. Both misclassified images (i.e. Fig. 6f and g) are visually alike *Williamson turn* and could be the possible reason for misjudgment. A lack of sufficient training data is a prevalent and considerable challenge in machine learning models. However, the model performed well in classifying predominantly different images (i.e. Fig. 6e and h).

The image recognition model can function as an additional tool in the proposed Semi-Automated Performance Evaluation System (SAPES) architecture in nautical training by Kim et al. [25]. Building on the comprehensive framework proposed by Munim et al. [26] for incorporating emerging technologies into maritime simulator training, the architecture framework in this study serves as a data analytics tool that uses image data to improve the evaluation of navigation performance. Figure 7 presents an integration architecture for the LAD component based on the findings of this study. Visual ship trajectory data from ECDIS can be obtained by taking screenshots during the exercise. The CNN-DNN algorithm will then analyze these images, and the results can be presented in various formats to assess a trainee's performance. This data can be displayed on multiple devices via, Application Programming Interface (API) integration, as needed.

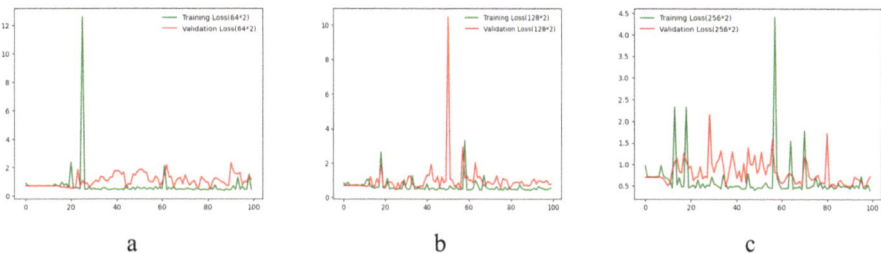

Fig. 4. Training and validation loss.

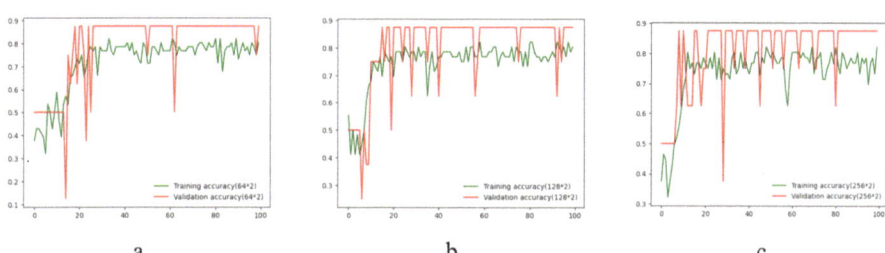

Fig. 5. Training and validation accuracy.

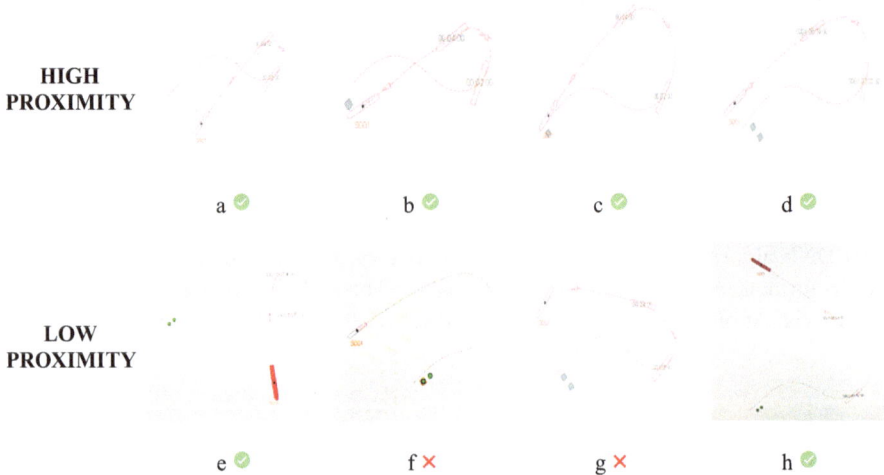

Fig. 6. Test images with labels (*Green* ✓: Correctly classified, *Red* ✗: Incorrectly classified)

Fig. 7. Proposed LAD component integration architecture for nautical simulation exercise

5 Conclusions

This study investigated the application of the image classification approach in performing simulator training of *Williamson turn*. The model was trained using a dataset of 64 images. A CNN-DNN model with different densities of neurons/ units were tested to generate the model. A model with 64 neuron layers was ultimately chosen due to lower training time and reasonable accuracy. Due to the limited size of samples, a dropout layer was used during model building to prevent overfitting [27]. The trained model misclassified two of the images among the test set of eight images. A larger dataset and refined neural network may improve the model's accuracy. The model can empower nautical students to independently practice and master manoeuvres such as *Williamson turn*. Despite the straightforward nature of *Williamson turn* instructions, students frequently deviated from MOB buoy. This was primarily attributed to errors in heading calculations and misjudgment of the ship's rate of turn. Students can review manoeuvre instructions and practice them on desktop simulators during their free time or without the presence of an instructor. They can upload images or screenshots of their exercise and run the image recognition model to examine their performance. It opens the possibility of personalized learning where a nautical student can learn simple manoeuvres like *Williamson turn.*

The *Williamson turn* manoeuvre was chosen for this study due to its widespread use and effectiveness in low visibility [28]. In this study, only ECDIS image data was used for evaluating one of many manoeuvring skills. Future studies can integrate simulator log data with image data to analyse student performance in simulator training. Although multi-faceted data analysis in simulator training evaluation is demanding, it promises to significantly enhance the depth and accuracy of descriptive and predictive insights into student learning. Generative AI models may be useful in offering training to individual student needs by pinpointing specific weaknesses identified through an integrated approach of analysis.

Acknowledgments. The authors would like to acknowledge contributions to simulator instructors at UiT for their support while performing the experiments. This study was funded by i-MASTER project (grant agreement No 101060107) under the European Union's Horizon Europe research and innovation programme.

Disclosure of Interests. The authors have no competing interests to declare that are relevant to the content of this article.

References

1. Domeh, V., et al.: Risk analysis of man overboard scenario in a small fishing vessel. Ocean Eng. **229**, 108979 (2021)
2. Pitman, S.J., Wright, M., Hocken, R.: An analysis of lifejacket wear, environmental factors, and casualty activity on marine accident fatality rates. Saf. Sci. **111**, 234–242 (2019)
3. Huang, P.-F., et al.: The design and development of man overboard alarm and rescue terminal. J. Discrete Math. Sci. Cryptography **19**(3), 649–661 (2016)
4. Örtlund, E., Larsson, M.: Man Overboard detecting systems based on wireless technology (2018)
5. Zhang, Y., Tao, Q., Yin, Y.: A lightweight man-overboard detection and tracking model using aerial images for maritime search and rescue. Remote Sens. (Basel, Switzerland) **16**(1), 165 (2024)
6. Su, Z., et al.: Serviceability of the IAMSAR standard man overboard recovery maneuvers: a case-study of full-scale sea trials. Appl. Ocean Res. **114**, 102782 (2021)
7. International Maritime, O., International Civil Aviation, O.: IAMSAR manual : international aeronautical and maritime search and rescue manual: vol. 3: Mobile facilities. 2010 edn. International Maritime Organization, London (2010)
8. Kobayashi, H.: Use of simulators in assessment, learning and teaching of mariners. WMU J. Marit. Aff. **4**(1), 57–75 (2005)
9. Rhienmora, P., et al.: Intelligent dental training simulator with objective skill assessment and feedback. Artif. Intell. Med. **52**(2), 115–121 (2011)
10. Sam, O., et al.: Objective and intelligent training assessment package for maritime training in simulator. J. Phys.: Conf. Ser, **2311**(1), 12014 (2022)
11. Munim, Z.H., Kim, T.E.: A review of learning analytics dashboard and a novel application in maritime simulator training. AHFE (2023)

12. Hjelmervik, K., Nazir, S., Myhrvold, A.: Simulator training for maritime complex tasks: an experimental study. WMU J. Marit. Aff. **17**(1), 17–30 (2018)
13. Hareide, O.S., Ostnes, R.: Maritime usability study by analysing eye tracking data. J. Navigation **70**(5), 927–943 (2017)
14. Atik, O.: Eye tracking for assessment of situational awareness in bridge resource management training. J. Eye Mov. Res. **12**(3) (2019)
15. Aljohani, N.R., et al.: An integrated framework for course adapted student learning analytics dashboard. Comput. Hum. Behav. **92**, 679–690 (2019)
16. Kim, I., Chae, C., Lee, S.: Simulation study of the IAMSAR standard recovery maneuvers for the improvement of serviceability. J. Marine Sci. Eng. **8**(6), 445 (2020)
17. Data of the i-MASTER project: a novel initiative in maritime education and training experience. 2023, UiT The Arctic University of Norway (UiT) https://dataverse.no/dataverse/uit: DataverseNO
18. Keysers, D., et al.: Deformation models for image recognition. IEEE Trans. Pattern Anal. Mach. Intell. **29**(8), 1422–1435 (2007)
19. Khellal, A., Ma, H., Fei, Q.: Convolutional neural network based on extreme learning machine for maritime ships recognition in infrared images. Sensors (Basel) **18**(5), 1490 (2018)
20. Richards, D.R., Tunçer, B.: Using image recognition to automate assessment of cultural ecosystem services from social media photographs. Ecosyst. Serv. **31**, 318–325 (2018)
21. Lou, G., Shi, H.: Face image recognition based on convolutional neural network. China Commun. **17**(2), 117–124 (2020)
22. Zhang, M.M., et al.: VAIS: A dataset for recognizing maritime imagery in the visible and infrared spectrums. The Institute of Electrical and Electronics Engineers, Inc. (IEEE), Piscataway
23. Russakovsky, O., et al.: ImageNet large scale visual recognition challenge. Int. J. Comput. Vision **115**(3), 211–252 (2015)
24. Krizhevsky, A., Sutskever, I., Hinton, G.: ImageNet classification with deep convolutional neural networks. Commun. ACM **60**(6), 84–90 (2017)
25. Kim, T.-E., Munim, Z.H., Schramm, H.-J.: Exploring alternative performance evaluation method in nautical simulations. Train. Educ. Learn. Sci. **109**(109) (2023)
26. Munim, Z.H., et al.: Scenario design, data measurement, and analysis approaches in maritime simulator training: a systematic review. in international conference in methodologies and intelligent systems for technology enhanced learning. Springer Nature Switzerland (2023)
27. Srivastava, N., et al.: Dropout: a simple way to prevent neural networks from overfitting. J. Mach. Learn. Res. **15**(1), 1929–1958 (2014)
28. Gil, M., Formela, K., Śniegocki, M.: Comparison of the efficiency of williamson and anderson turn manoeuvre. TransNav: Int. J. Marine Navig. Safety Sea Transport. **9**(4) (2015)

Sensitivity of Predictive Performance Assessment Accuracy in Varying k-fold Cross Validation

Fabian Kjeldsberg[1], Ziaul Haque Munim[1(✉)], Morten Bustgaard[1], Sahil Bhagat[2], Emilia Lindroos[3], and Per Haavardtun[1]

[1] Faculty of Technology, Natural and Maritime Sciences, University of South-Eastern Norway, Horten, Norway
`ziaul.h.munim@usn.no`
[2] Department of Technology and Safety, University of Tromsø (UiT), The Arctic University of Norway, Tromsø, Norway
[3] Faculty of Technology and Seafaring, Novia University of Applied Sciences, Turku, Finland

Abstract. In machine learning (ML) applications, cross-validation (CV) allows greater generalizability of a trained algorithm over out-of-sample or new data. This study explores the accuracy of trained ML algorithms in predicting student performance in a maritime simulator exercise scenario in four different k-fold CVs. Three, five, eight, and ten-fold CVs were trained using a cloud-ML platform. Three top-performing ML algorithms were evaluated considering log loss, accuracy, and area under the curve (AUC). The results indicate higher predictive accuracy with increasing k in CV folds. Considering the trade-off between prediction accuracy and the time required to predict every 1000 observations, using the five-fold CV in predictive learning analytics appears optimal in the explored simulation training scenario. Prediction explanations of five-fold CV are reported.

Keywords: machine learning · cross-validation · navigation simulator · maritime training · learning analytics

1 Introduction

Learning analytics refers to "the measurement, collection, analysis and reporting of data about learners and their contexts, for purposes of understanding and optimizing learning and the environments in which it occurs" [1]. Implementation of learning analytics typically include four phases: data generation, data tracking, data analysis, and action [2]. While the majority of extant research focuses on learning analytics in online courses and platforms [3], its application in complex simulation training contexts is limited. In maritime simulator training, data can be generated from simulator logs, tracked for each student, analysed for patterns, and implemented for prediction of performance [4].

© The Author(s) 2025
T. E. Kim et al. (Eds.): MIS4TEL 2024, LNNS 1274, pp. 71–82, 2025.
https://doi.org/10.1007/978-3-031-84170-5_7

Learning analytics is the core of Learning Analytics Dashboard (LAD), which are mainly 3 types depending on the degree of data analysis and action: descriptive, predictive, and prescriptive [5]. In predictive elements of LADs, insights are usually generated using machine learning (ML) algorithms [6]. In ML applications, while training algorithms, it is a common practice to divide the available dataset into training and test samples. Typically, 80-20 or 70-30 training-test split is used. Akay experimented with various train-test splits on a breast cancer dataset for a binary classification model [7]. It has become common to consider a validation split too, for example, 60-20-20 split where 60% data is used in training, 20% in validation, and 20% in testing. Ran et al. used a primary and external validation dataset to test a 3D deep-learning system to detect glaucomatous optic neuropathy [8]. Such split of available data allows for testing the performance of trained algorithms over varying data points, which enhances the generalizability of the algorithms to new unseen data. This reduces the overfitting or underfitting [9] issue in ML applications. To further enhance generalizability of the trained ML models, the cross-validation (CV) approach is widely used. ML algorithms are trained over multiple training samples from the same dataset and validated over multiple validation samples. The average model performance score of the multiple training and validation samples is benchmarked to decide on the best-suited ML algorithm.

There are several CV approaches in ML applications. K-fold and stratified CV are most common [10]. Among the various types of CV, *K*-fold CV is extensively used in selection and error estimation of classification models [11, 12]. Essentially, cross validation involves dividing the data into 'k' subsets or folds. Model training and validation is carried out 'k' times. In each iteration, one of the 'k' subsets is used as the validation set, and the remaining 'k-1' subsets are used for training. The 'k' establishes how many folds the dataset is partitioned into. By increasing the number of folds, the number of subsets for training rises correspondingly. The error metrics from each iteration are compiled and the mean value is calculated to determine an overall average error metric [13]. As an example, the structure of five-fold cross-validation is graphically represented in Fig. 1.

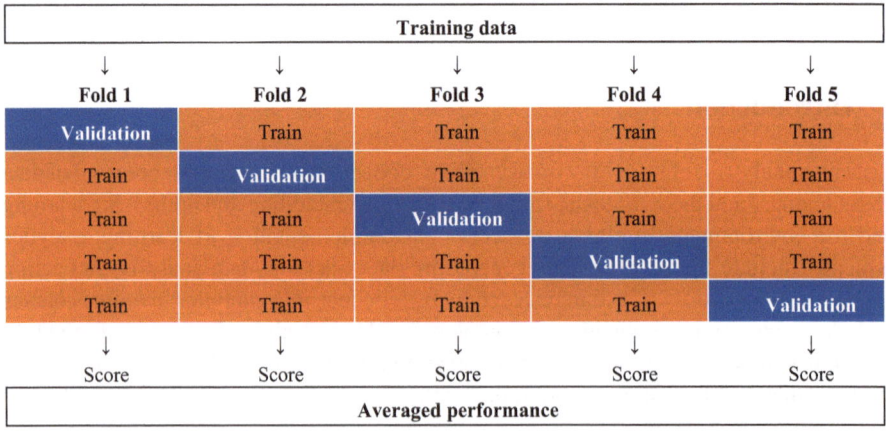

Fig. 1. Five-fold cross validation process illustrated

In an effort to developing a learning analytics dashboard (LAD) in the context of maritime navigation simulation training, Munim et al., [14] explored a range of ML algorithms. A desktop navigation scenario, Williamson turn, was designed in Kongsberg's k-sim simulator. Simulator log data at 1-s interval was extracted, and instructors evaluated the performance of students on an ordinal scale: Good, Satisfactory, and Needs Improvement. 13 extracted simulator log data parameters were used as input to predict the evaluated student performance. This study investigates how the performance of the top three algorithms in the predictive performance assessment of maritime students changes in different k's in k-fold CV.

2 Simulation Training Scenario

The *Williamson turn* is a maritime navigational procedure executed during search and rescue operations, specifically in instances such as a Man Overboard (MOB) event [15]. This maneuver involves an initial hard starboard (right) turn, followed by a portside (left) U-turn. The purpose of this procedure is to reverse the vessel's course in a precise and controlled manner, enabling it to return to the location of the MOB incident. The training could be conducted for a ballast or a loaded ship. Data used in this study are from a loaded ship exercise.

In this study, eight students participated in the simulation training exercise. All participants were informed about the manoeuvre, and consent was obtained before data collection. The desktop simulation lab where the training was conducted has 12 bridges, hence, four bridges were empty during the training. Figure 2 presents an illustration of the exercise from the instructors' dashboard.

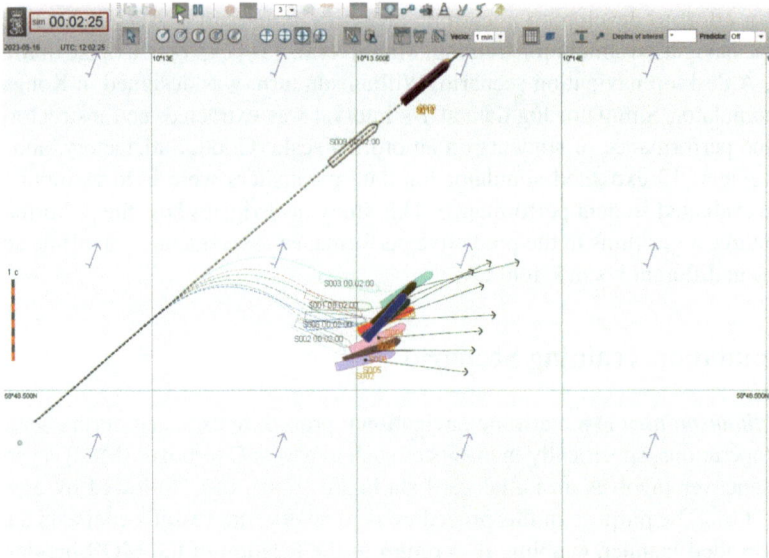

(a) Recording snapshot at 02:55 minutes in the exercise

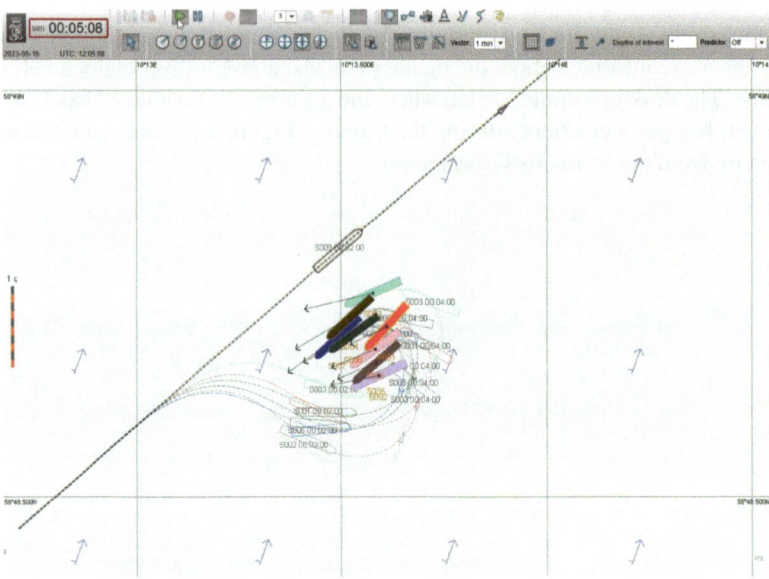

(b) Recording snapshot at 05:08 minutes in the exercise

Fig. 2. Snapshot from the Williamson Turn exercise

3 Methodology

3.1 Simulator Log Data

We utilize simulator log data from eight university-level student exercises each performing a *Williamson turn* in loaded vessel conditions retrieved from desktop simulators. Data of 13 variables (Table 1) with one-second frequency are collected and comprise 3432 data rows in total, 429 rows per student. Data is divided into training-validation-test subsamples. Data of seven students are used in training and validation of ML algorithms, that is, 3003 rows of data (87.50%). The rest of the 12.50% data, corresponding to log data of one student, was kept for test or holdout samples.

Table 1. Variables in Dataset

No.	Variables	No.	
1	Main propeller revolutions	8	Latitude
2	Speed over ground	9	Longitude
3	Course	10	Distance sailed from Log sensor
4	Wind sensor 1 direction	11	Main Becker rudder angle
5	Wind sensor 1 speed	12	Heading
6	List	13	Course order
7	Rate of turn		

3.2 Automated Machine Learning

The training dataset is analyzed with three-, five-, eight-, and ten-fold CV to measure and determine the optimal number of CV folds for predicting student performance in a maritime simulator exercise scenario.

For analysis, ML algorithms are trained through the cloud-ML platform *DataRobot* (https://www.datarobot.com/), an Automated machine learning (AutoML) platform. AutoML simplifies the application of ML by taking care of repetitive tasks in the process, and allows the possibility to train a large number of relevant ML algorithms by automating feature engineering, hyperparameter optimization, validation, and model selection [16, 17]. A total of 95 ML algorithms are trained and tested in this study, 23 in three-fold and five-fold, 24 in eight-fold, and 25 in ten-fold. While training algorithms, DataRobot increases the training sample raising from 16% random sample to 64% in three iterations. Hence, in the first training iteration, 16% of the 3003 observations are used in training, then 32% and 64% of 3003 observations in the 2nd and 3rd iterations, respectively. The top-performing algorithms are determined based on prediction error metrics scores.

4 Results

4.1 Prediction Error Metrics

The results from AutoML indicate that the eXtreme Gradient Boosted Trees, Light Gradient Boosted Trees, and Random Forest Classifier algorithms are consistently the best-performing models across the CV configurations explored in this study. Hence, the two primary ML architectures utilized in this study are gradient-boosted trees and random forest. Data shown in Figs. 3, 4, and 5 present three error metrics values (i.e., loss, accuracy and Area Under Curve (AUC) respectively) from k-fold cross-validation. In terms of LogLoss (Fig. 3), eXtreme Gradient Boosted Trees' performance is the weakest in the five-fold configuration. Random Forest records continually improved performance as folds increase, and Light Gradient Boosted Machine's performance exhibits a similar pattern. While all models showed variations in the validation set with higher fold, the random forest classifier has notably improved than the other two.

In terms of accuracy (Fig. 4), Light Gradient Boosted Machine exhibits better performance in five, eight, and ten-fold CVs. The AUC metric (Fig. 5) reveals identical results to accuracy. There seems to be a trend that increasing the number of folds generally improves the performance metrics scores for all models. Depending on the model, different fold configurations seem to be optimal. Although the best accuracy metrics scores are recorded on the eight and ten-fold CVs, there is only a slight improvement from the five-fold configuration.

	XGBoost	Light GBM	Random Forest			
3-fold	0.2163	0.2753	0.2924	Validation	0.30	High
	0.2259	0.2853	0.2955	CV	0.29	
	0.2082	0.2707	0.2866	Holdout	0.28	
5-fold	0.2525	0.2177	0.2487	Validation	0.27	
	0.2582	0.2242	0.2625	CV	0.26	
	0.2497	0.2343	0.2733	Holdout	0.25	Medium
8-fold	0.2086	0.2014	0.2126	Validation	0.24	
	0.2495	0.2217	0.2512	CV	0.23	
	0.2363	0.2353	0.2512	Holdout	0.22	
10-fold	0.2050	0.2029	0.1985	Validation	0.21	
	0.2448	0.2231	0.2467	CV	0.20	
	0.2421	0.2275	0.2473	Holdout	0.19	Low

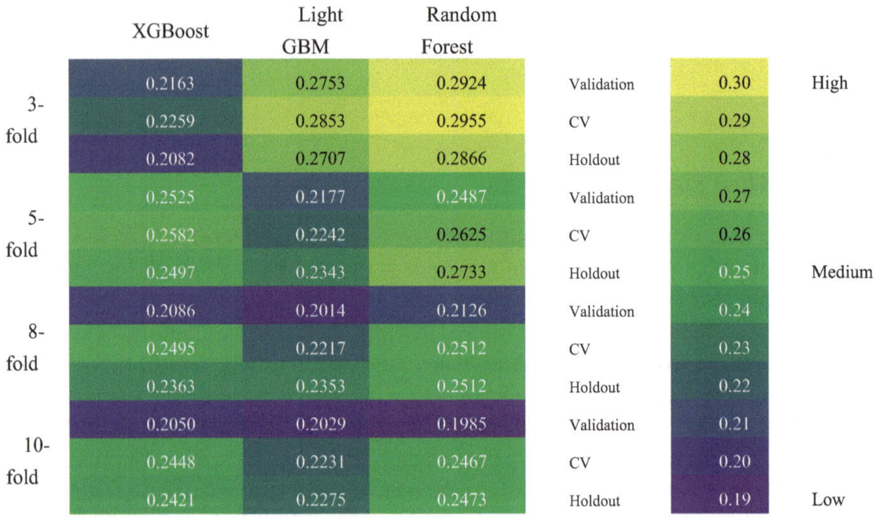

Fig. 3. LogLoss for different models and fold configurations

4.2 Prediction Computation Time

Computational times vary depending on model and fold configuration (Fig. 6). XGBoost is the fastest model on the three-fold configuration. At the same time, Light GBM is

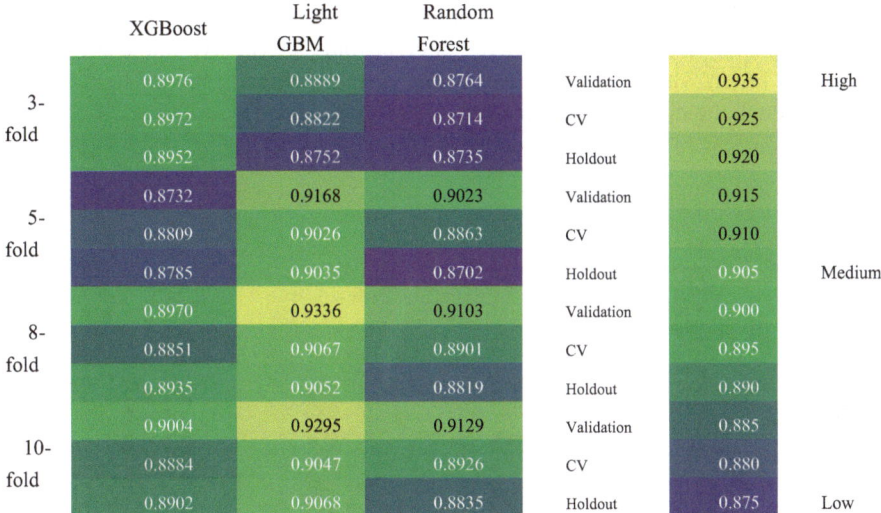

Fig. 4. Accuracy for different models and fold configurations

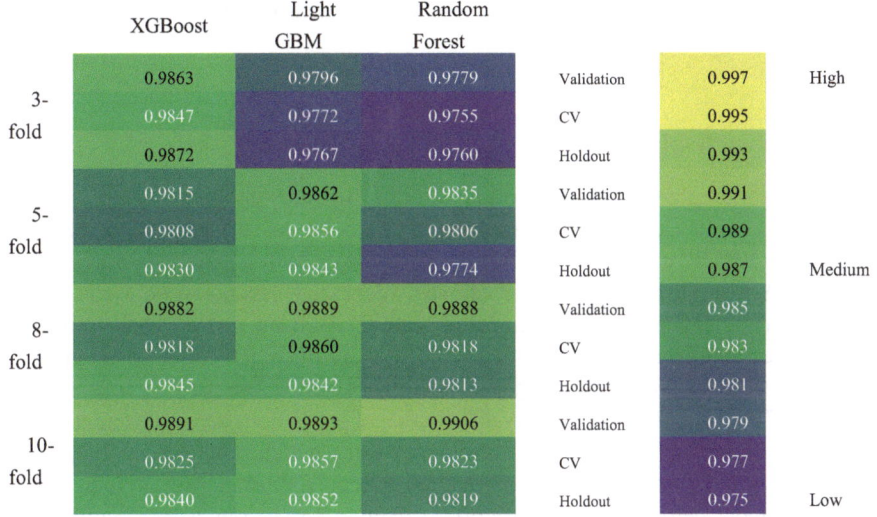

Fig. 5. AUC for different models and fold configurations

faster on five-, eight-, and ten-fold configurations. Generally, all models perform faster in the five-fold iteration than three-fold except for the XGBoost. XGBoost's computational time exhibits a steady increase with the increase in fold configuration (Fig. 6). However, for Light GBM and Random Forest, the five-fold configuration seems to be less computationally heavy than any other configurations. Random Forest is the slowest model across all different CV configurations. Random Forest records the highest computational time on the three- and ten-fold configurations.

The optimal balance between error metrics scores and computational times in predictive learning analytics is observed in five-, or eight-fold CV. The five-fold configuration is typically faster, whereas the eight-fold setup slightly outperforms in terms of error metrics, though the difference is minimal.

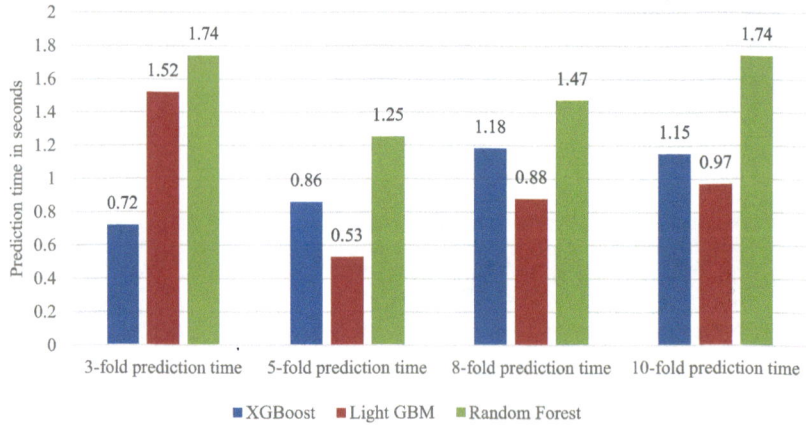

Fig. 6. Prediction Time for All Models Across Different Fold-Configurations

4.3 Prediction Explanation

The prediction explanations for five-fold Light GBM are reported in Fig. 7a–c. The explanations are produced using DataRobot's eXemplar-based Explanations of Model Predictions (XEMP) methodology. Only the top five explanatory parameters are reported in Fig. 7. Across all three prediction classes, "geometry" seems to be a consistently impactful feature, which is an engineered feature based on Latitude and Longitude. The specific values that lead to each classification are different. Here, we interpret the "Satisfactory" class predictions (Fig. 7a), and others can be interpreted similarly.

Figure 7 represents the probability distribution of predictions across different classes. It shows two thresholds: (1) a low threshold (left side), which is marked in blue, and (2) a high threshold (right side), which is marked in red. The low threshold indicates the probability score below which a prediction is likely not to belong to the "Satisfactory" class, and the high threshold is the score above which the model is quite confident that the prediction should be classified as "Satisfactory."

The prediction in Fig. 7a has a high confidence score (0.998201), which is higher than the high threshold and close to 1, suggesting that the model is very confident of this particular classification. The Fig. 7a is for the "Satisfactory" class, which indicates that the prediction we are looking at has been classified as "Satisfactory" by the model. Row ID is a unique identifier for the data row in the dataset that this prediction corresponds to (i.e., Row ID: 2491).

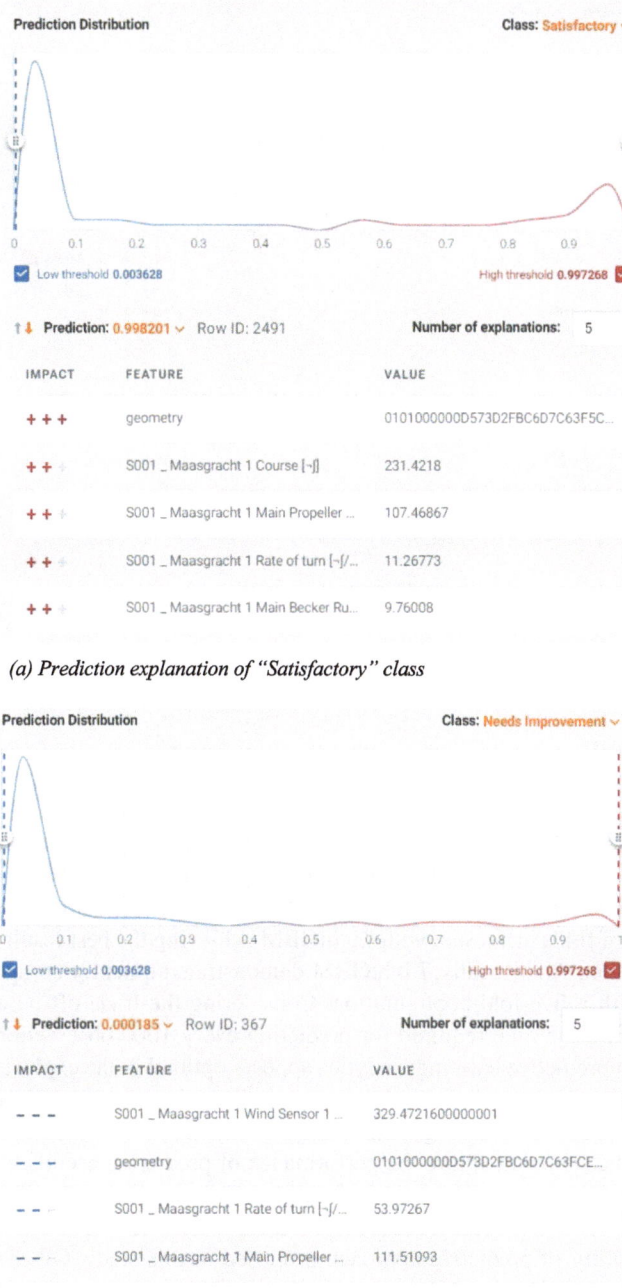

(a) Prediction explanation of "Satisfactory" class

(b) Prediction explanation of "Needs Improvement" class

Fig. 7. Prediction Explanation of 5-fold

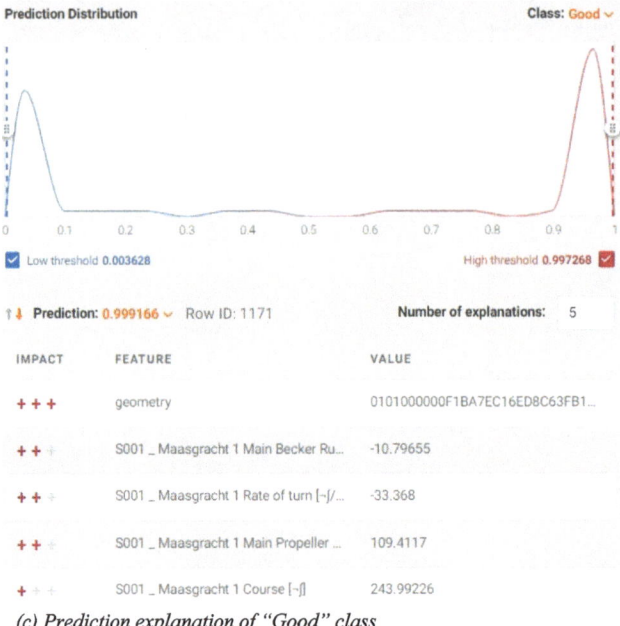

(c) Prediction explanation of "Good" class

Fig. 7. (*continued*)

5   Conclusion

This study investigates the sensitivity of prediction accuracy in simulator training performance assessment with changing CV folds. Error metric scores and computational times of three ML algorithms on three-, five-, eight-, and ten-fold CVs on a dataset compiled from simulator log data are explored. The accuracy for all models enhanced as the number of folds increased, with LightGBM achieving the best results in an eight-fold configuration. Additionally, LightGBM demonstrated quickest computation time, particularly with a five-fold configuration. Considering the trade-off between prediction accuracy and the time required for predicting every 1000 observations, using the five-fold CV in predictive learning analytics appears optimal in the explored simulation training scenario. Such predictive performance assessment can be utilized in a LAD, particularly in a maritime training context.

Future studies should explore the performance of prediction accuracy of simulator log data from a larger sample of students performing the the same *Williamson turn* under varying k-fold CVs. Larger and more balanced datasets will be able to improve the generalizability of predictive approach proposed in this study. Other types of CV folds such as stratified folds may produce further insights into model performances for learning analytics.

Acknowledgments. The authors thank the students who participated in the simulator training exercise. This research is funded by the European Union's Horizon Europe research and innovation programme under grant agreement No 101060107.

References

1. Siemens, G., Long, P.: Penetrating the fog: analytics in learning and education. Educ. Rev. **46**(5), 30 (2011)
2. Wong, B.T., Li, K.C.: A review of learning analytics intervention in higher education (2011–2018). J. Comput. Educ. **7**(1), 7–28 (2020)
3. Knobbout, J., Van Der Stappen, E.: Where is the learning in learning analytics? A systematic literature review on the operationalization of learning-related constructs in the evaluation of learning analytics interventions. IEEE Trans. Learn. Technol. **13**(3), 631–645 (2020)
4. Munim, Z.H., Krabbel, H.L.S., Haavardtun, P., Kim, T.-E., Bustgaard, M., Thorvaldsen, H.: Scenario design, data measurement approaches, and analysis methods in maritime simulator training: A systematic review. In: Methodologies and Intelligent Systems for Technology Enhanced Learning, Guimarães, Portugal (2023)
5. Susnjak, T., Ramaswami, G.S., Mathrani, A.: Learning analytics dashboard: a tool for providing actionable insights to learners. Int. J. Educ. Technol. High. Educ. **19**(1), Art. no. 1 (2022). https://doi.org/10.1186/s41239-021-00313-7
6. Munim, Z.H., Kim, T.-E.: A review of learning analytics dashboard and a novel application in maritime simulator training. In: Proceedings of the AHFE Conference (2023)
7. Akay, M.F.: Support vector machines combined with feature selection for breast cancer diagnosis. Expert Syst. Appl. **36**(2), 3240–3247 (2009)
8. Ran, A.R., et al.: Detection of glaucomatous optic neuropathy with spectral-domain optical coherence tomography: a retrospective training and validation deep-learning analysis. Lancet Digit. Health **1**(4), e172–e182 (2019)
9. Burzykowski, T., Geubbelmans, M., Rousseau, A.-J., Valkenborg, D.: Validation of machine learning algorithms. Am. J. Orthod. Dentofacial Orthop. **164**(2), 295–297 (2023)
10. Krstajic, D., Buturovic, L.J., Leahy, D.E., Thomas, S.: Cross-validation pitfalls when selecting and assessing regression and classification models. J. Cheminform. **6**(1), 1–15 (2014)
11. Anguita, D., Ghelardoni, L., Ghio, A., Oneto, L., Ridella, S.: The "K" in K-fold cross validation. Comput. Intell. (2012)
12. Fushiki, T.: Estimation of prediction error by using K-fold cross-validation. Stat. Comput. **21**(2), 137–146 (2011). https://doi.org/10.1007/s11222-009-9153-8
13. Jack Feng, C.-X., Yu, Z.-G.S., Kingi, U., Pervaiz Baig, M.: Threefold vs. fivefold cross validation in one-hidden-layer and two-hidden-layer predictive neural network modeling of machining surface roughness data. J. Manuf. Syst. **24**(2), 93–107 (2005). https://doi.org/10.1016/S0278-6125(05)80010-X
14. Munim, Z.H., et al.: Performance assessment in a maritime rescue simulation training scenario using machine learning. Unpublished (2024)
15. Su, Z., Liu, X., Zheng, G., Zhou, K., Gao, S.: Serviceability of the IAMSAR standard man overboard recovery maneuvers: a case-study of full-scale sea trials. Appl. Ocean Res. **114**, 102782 (2021)
16. He, X., Zhao, K., Chu, X.: AutoML: a survey of the state-of-the-art. Knowl.-Based Syst. **212**, 106622 (2021). https://doi.org/10.1016/j.knosys.2020.106622
17. Karmaker, S.K., Hassan, M.M., Smith, M.J., Xu, L., Zhai, C., Veeramachaneni, K.: Automl to date and beyond: challenges and opportunities. ACM Comput. Surv. CSUR **54**(8), 1–36 (2021)

Using Emerging Technologies for Bridging Computational Thinking and Artificial Intelligence in Swedish Classrooms: Empirical Insights and Pedagogical Implications

Rafael Zerega(✉) ⓘ, Johanna Velander(✉) ⓘ, and Marcelo Milrad(✉) ⓘ

Department of Computer Science and Media Technology, Linnaeus University, Växjö, Sweden
{rafael.zerega,johanna.velander,marcelo.milrad}@lnu.se

Abstract. The increasing use and application of Artificial Intelligence (AI) across various sectors in society, combined with the growing adoption of emerging technologies, emphasizes the urgent necessity to develop digital competencies in K-12 education, encompassing both digital literacy and AI literacy. Computational thinking (CT) has been recently incorporated into the educational curricula of several countries as it is regarded as an effective approach to improving problem-solving and programming skills. AI literacy, however, has only recently started to gain attention and make its way into K-12 teaching. Few studies to date have investigated to what extent having a technical understanding of AI methods and techniques can contribute to actual AI literacy thus finding effective ways to integrate this knowledge into policy and curricula. To address this problem and contribute with empirical insights, we designed and conducted a study identifying the challenges and the opportunities teaching basic concepts of AI to a class of middle school students. During a workshop series, an entire class of 8th graders engaged in hands-on activities related to CT and AI with a particular focus on machine learning. We identify a lack of prerequisite skills that impact students' ability to engage in contextual AI-related activities due, at least partly, to poor programming skills. This, in turn, seems detrimental to gaining transferable skills in machine learning which are important given the interdisciplinary nature and broad application of these AI methods. We identify CT and programming as prerequisite skills necessary to achieve a deeper understanding of AI technology and its implications for individuals and society.

Keywords: Computational Thinking · Programming · Artificial Intelligence · Machine Learning · AI literacy · Digital Competencies · K-12 education

1 Introduction

As emerging technologies like AI and educational robotics become increasingly prevalent, there is a pressing demand for educational curricula in K-12 settings to incorporate relevant teaching methods and content. Many countries are already providing education aimed at fostering the development of digital competencies as part of their strategies

© The Author(s) 2025
T. E. Kim et al. (Eds.): MIS4TEL 2024, LNNS 1274, pp. 83–97, 2025.
https://doi.org/10.1007/978-3-031-84170-5_8

towards digitalization in education. Although programming has been present in K-12 education for quite some time, AI-related education is still trying to find its place in this context (Touretzky & Gardner-Mccune 2019). One of the approaches to introducing programming to elementary schools has been through Computational Thinking (CT), which has been incorporated into the educational curricula in several countries (Grover & Pea 2018; Kafai & Burke, 2017). As for the increasing need for education on AI in schools, different organizations have issued recommendations guiding countries worldwide to design their educational programs considering AI technology. For example, UNESCO has issued recommendations for AI curriculum development for K-12 (Miao et al. 2021), likewise, in October 2022 the European Union updated its digital competence framework Dig Comp 2.2 to include aspects of data and AI (Voukari et al. 2022).

Introducing AI concepts to students unfamiliar with the topic poses a significant challenge in the classroom. AI literacy in an educational context requires an interdisciplinary approach covering fundamental concepts of AI technology, its integration in applications, and its societal consequences. To better understand this process, we conducted a small exploratory study to get some insights regarding middle school students' readiness to engage in AI-related education, particularly ML and, to identify possible challenges that may arise along the process. To reach these goals the following research questions were defined: (1) *What challenges can be identified when introducing AI and ML education to middle school students?* and (2) *What role does CT (programming) play as a prerequisite knowledge to engage meaningfully in AI literacy education?*

To explore these questions, we designed and organized a series of workshops centered on CT and AI for middle school students in southern Sweden, forming the basis of our exploratory study. Throughout these workshops and using different tools and methods, the students reviewed basic concepts of rule-based programming and were introduced to fundamental concepts of AI technology, particularly ML. By engaging the students in these activities, we aim at gaining some insights regarding the challenges of introducing students to the topic of AI technologies while analysing the possible connection of this subject with CT. The remainder of the paper is structured as follows: the next section provides the theoretical perspectives that guided the design of this study, namely computational thinking, and AI literacy. The proceeding section describes the rationale for the workshop design and its implementation followed by the methods chosen for analysing the data we collected. We then report the obtained results and finalize this paper by discussing the implications of these outcomes. The paper ends by offering insights and reflections, along with recommendations for future research efforts connected to this study.

2 Theoretical Perspectives

In this section, we describe two concepts that are relevant to the workshop design and that serve as the basis for analysis of the learning process, namely computational thinking and AI literacy. It is important to clarify that these concepts are described in the context of K-12 education.

2.1 Computational Thinking in K-12

Computational thinking (CT) is an approach to problem-solving that has been gradually introduced in the curricula for K-12 education in several countries around the world and it is regarded as an important skill for problem-solving (Bocconi 2016; Grover et al. 2013). Countries like Australia, the USA, Finland, Estonia, and the UK, to mention just a few, have introduced CT in their K-12 curricula to develop programming skills and other digital competencies among students (Grover & Pea 2018). However, CT in the context of K-12 education remains a vaguely defined concept and thus, it has been relatively difficult to agree on how to impart CT knowledge (Grover & Pea 2013; Selby & Woolland 2013). Some scholars have defined ways of categorizing what are the core concepts of CT as an approach to problem-solving, ranging from concepts that are exclusively connected to the process of programming and building algorithms, to other higher-level aspects of CT such as the role of collaboration and the importance of questioning (Brennan & Resnick 2012; Grover & Pea 2018). Although it has been suggested that in its current form CT does not include relevant aspects of AI and that it should be upgraded to better reflect AI technology (Dohn et al. 2022; Tedre et al. 2021), some concepts of classic CT are still relevant for understanding some of the underlying approaches and methods of AI technology (Heintz 2022; Zerega & Milrad 2023). Throughout this article, we will refer to CT and rule-based programming interchangeably.

2.2 AI Literacy in K-12

Literacy is a term that has long been used to describe abilities to make meaning out of and express oneself through written text (Knobel & Lankshear 2014). As text is nowadays communicated and distributed through more means than printed static documents, new literacies accounting for such practices have emerged (Knobel & Lankshear 2014). Rapid development and broad uptake of digital technologies have sparked a growing interest in digital literacy (Merchant 2021) which has lately expanded to include AI literacy (Long & Magerko 2020) and data literacy (Wolff et al. 2016) to account for data-driven practices commonly embedded in digital solutions and applications (Velander et al. 2023). Digital and data-driven practices such as ML rely on data representing individuals and objects by proxy, influencing decision-making processes, recommendations, and predictions used in most domains and parts of society (Merchant 2021). As such it has implications for social, political, and democratic engagement and thus calls for a critical perspective on literacy rather than a more traditional or functional literacy. Functional literacy prepares individuals with relevant skills and knowledge to be part of a future life and workforce, critical literacy also provides prerequisites for civic engagement and social empowerment (Giroux 1998). Recent reviews on the conceptualization of AI literacy, however, reveal a prominent focus on functional literacy and largely leave out critical perspectives on AI concepts, applications, and ethical implications that facilitate reflections on past, present, and future use of technologies (Velander et al., in press). AI literacy is currently being promoted to be included in policy and curricula in K-12 education, for example, the EU has recently extended its digital competence framework, DigComp 2.2. (Vourkari et al. 2023) to account for AI and data literacy.

3   Methodological Approach

This section presents important aspects regarding the methodological approaches used in the design and implementation of the workshops.

3.1   Informing the Design of the Workshops

The proposed workshop activities were designed considering specific pedagogical methods, characteristics of the group of students that participated in the workshops and the learning objectives that we defined based on existing frameworks about CT and AI in K-12 and study plans for developing basic AI knowledge.

Pedagogical Methods. We based the design of the workshop activities on two main pedagogical methods, namely *challenge-based learning*, and *inquiry-based learning*, both with a *constructionist approach*. We used challenge-based learning for hands-on activities where students were required to devise and implement a working solution to address a specific problem or challenge (Conde et al. 2019). In addition, we used inquiry-based learning as the pedagogical method for those hands-on activities where students had to try out different tools and systems based on AI technology as we wanted to engage students in reflecting on the main characteristics of this technology and the potential of systems based on AI and ML (Pedaste et al. 2015). The written exercises aimed at getting students to reflect on how AI technology works and how it compares to traditional rule-based programming were also based on inquiry-based learning.

Settings and Participants. The workshops in this study were conducted during the autumn semester of 2023/2024 with seventeen 8th-grade students aged 14–15. These workshop sessions took place in a public school run by a municipality in southern Sweden, and it has a specific orientation towards natural sciences and technology. It is important to mention that these workshops took place as part of their curricular schedule and were agreed upon with their head teachers and the students' parents who were sent a letter of consent where they were informed about the workshops that were going to take place, the topics to be covered and the type of data to be collected during the workshop sessions.

Pre-intervention Questionnaire to Determine Students' Previous Knowledge. Before the workshop sessions, the students had to fill in a questionnaire that provided us with relevant information regarding their previous knowledge of programming and AI. We used the information from this questionnaire to choose the type of activities and tools for the workshops.

The Swedish Curriculum to Determine Students' Previous Knowledge. In Swedish K-12 education both CT and AI literacy is only implicitly included under the umbrella term Adequate digital competence (Skolverket [the Swedish National Agency for Education] 2022). All teachers at all levels of K-12 education and all subject topics are responsible for providing the necessary education for students to become digitally competent. However, as the concept is vaguely defined and lacks specific assessment criteria it is open to interpretation by teachers and often must give way to defined and assessed teaching (Velander et al. 2023). We assumed students participating in this workshop series already possessed certain knowledge of basic programming, as according to the current Swedish curriculum, 8th-grade students have already received education in this topic. According to this curriculum (Skolverket 2018), the subjects of mathematics explicitly mention programming in the overall objectives and both topics of mathematics and technology further specify learning goals for programming for primary 1–3, 4–6, and 7–9.

Workshop Structure. We conducted five workshops for a total of 16 classroom hours. The activities focused on AI were loosely based on the MIT AI Ethics Curriculum (Payne 2019), and the AI literacy competency framework by Long and Magerko (2020), whereas the activities focusing on CT and programming were based on the CT framework from Brennan and Resnick (2012). The following is the general structure of each of the workshop sessions:

1. Brief keynote presentation of the main topics to be covered during the workshop session.
2. Hands-on activities on the main topics of the session using different tools and materials.
3. A moment for discussing the main topics covered in the workshop session.
4. Written exercise about the main concepts covered in the workshop session.

Table 1 below depicts the learning objectives for each hands-on activity, the description of the actual activity and the tool used to teach those learning objectives. Sessions 1, 2 and 5 focused on getting the students to test different systems that use AI technology and then engaging in reflective activities such as group discussion and answering written exercises. These activities were based on inquiry-based learning. Session 2 and 4 were based on challenge-based learning and they required the student to devise a working solution to address a specific challenge or problem. While in session 2 the solutions created by the students were based on traditional rule-based programming only, in session 4 the students had to devise a solution based on rule-based programming in combination with an ML model that students had to create and train.

3.2   Data Collected and Analysis Method

During the workshop series, we collected data of different types. The analysis will be based on qualitative content analysis. The data came from the following sources:

- **Screen recordings:** from the programming-related activities using block-based programming in sessions 2 and 4 (640 min of video footage).
- **Sound recordings:** from the group discussions during sessions 3, 4 and 5 (47 min of recorded audio).
- **Participant-generated content:** from written exercises regarding programming and AI from sessions 3, 4, and 5 (a total of 65 pages of text)
- **Observation and field notes:** Notes taken by the research team aiming at capturing students' attitudes and behaviors.

Qualitative Content Analysis. A large and varied amount of data was collected during the workshops. We used Qualitative Content Analysis as we wanted to understand students' progress. However, to avoid being constrained by the framework, our deductive

Table 1. Learning objectives for each hands-on activity, description of activity and tools used.

Session	Learning Objectives	Activity Description	Tools and Materials
S 1	Understand the potential of AI and ML to deliver natural human-computer interaction	Students get to interact with a robot that uses AI-based techniques for image and voice recognition	Misty Robotics
	Understand the effect training data has on the accuracy of a machine learning system. Human agency in curating datasets. Supervised ML = input – algorithm – output. Quality of training data affects accuracy of ML model (Payne 2019)	Image classification using Google Teachable Machine	Google Teachable Machine
S 2	Human role in training ML models AI learns from data Artificial vs human intelligence	Students train Misty the robot to recognize faces and react in a specific way upon recognition	Misty Robotics
	Applying CT concepts and practices based on the CT framework from Brennan & Resnick (2012)	Students program wheeled robots equipped with different sensors using block-based programming	Engino Robotic Platform
S 3	Basics of supervised (linear regression) and unsupervised (clustering) ML. Problem of classification in the supervised machine-learning context How quantity of training data affects the accuracy and robustness of a supervised machine-learning model (Payne 2019, p. 7)	Students collect data about their height and shoe size and import this dataset for use in the Python code. The features are plotted and using linear regression and K means clustering we visualize supervised and unsupervised ML	Python using: numpy, pandas, matplotlib, KMeans sklearn.cluster

(*continued*)

Table 1. (*continued*)

Session	Learning Objectives	Activity Description	Tools and Materials
S4	Understand the effect training data has on the accuracy of a machine learning system Applying CT concepts and practices based on the CT framework from Brennan & Resnick (2012) in combination with an ML model to create an application that can solve a specific problem or challenge	Students create and train ML models based on data they provided and test their accuracy (confidence level) Students export the models they created in ML for Kids and incorporate them in a Scratch project to create a working solution using block-based programming,	Scratch
S5	Interact with AI robots through vision and reflect on the potential role of robots and other AI-based systems in society	Students dance with Misty to see what body parts Misty recognizes and how it learns to mimic the dance moves of the student	Misty Robotics
	Reflecting on ethical issues regarding systems controlled by AI methods in our society	Students are presented with several moral dilemmas that self-driven cars may have to resolve where they must choose between two unfavourable outcomes (who gets to die and who gets to live)	The Moral Machine

approach was complemented with inductive coding using an unconstrained matrix (Elo and Kyngäs 2008) allowing us to expand and reflect on the existing framework. The following steps were taken during the analysis process: in the first step, one of the authors transcribed all data from audio recordings and the written material handed in by students on paper. Another author then translated the material from Swedish to English. In this way, both authors familiarize themselves with the data. The second step entailed deductively coding the data based on the framework by Long and Magerko (2020) and Brennan and Resnick (2012) whilst we iteratively built a matrix and inductively expanded it with additional categories as we reviewed the collected data. In the final step, both authors discussed the data labelling and resolved possible disagreements.

3.3 Frameworks for Analysis and Assessment

The data generated by the students during the five teaching interventions were analysed using two frameworks well-known in their respective areas. For analysing students' performance when engaging in rule-based programming we used the framework for assessing the development of CT by Brennan and Resnick (2012). This framework covers different dimensions of CT, some of them going beyond mere programming. In this study, we will focus mainly on the aspects of CT related to constructing algorithms and programming, referred to as CT concepts. As for the activities focused on AI- and ML-related topics we used the framework for AI literacy and competencies from Long and Magerko (2020). This framework enumerates a series of competencies regarding AI technology categorized into themes: *what is AI?, what can AI do?, how does AI work?, how should AI be used? and how is AI perceived?* In the next section, we present the results based on the data analysis.

4   Results

In this section, we delineate the outcomes derived from select workshop activities, specifically those focusing on the hands-on tasks and the written exercises. We first present results from the pre-intervention questionnaire to get insights regarding students' initial knowledge on programming and AI. We then present results from the rule-based programming activities. Then we present results from some of the AI-related activities and lastly, we present results from activities where students had to bring together rule-based programming and AI techniques. Afterward, we proceed to present the outcomes of certain AI-related activities, followed by a description of the results obtained from tasks that required students to integrate rule-based programming with AI techniques.

4.1   Results from the Pre-intervention Questionnaire *(N = 17)*

Some of the main results from this questionnaire can be summarized as follows:

- All 17 students report having received previous education in programming: (15) students had done block-based programming and (7) students had done textual programming.
- Most students (15) gain their knowledge about AI from social media and other incidental sources and only (3) from school (intentional source).
- Most students associate AI with robots (programmed or thinking) and chatbots, in particular ChatGPT. When asked what differentiates AI from other technology many (7) mention its ability to think and understand and some also mention that it can learn from experience or data (3). A few say it is smart (2) and has feelings (2).
- Three students have never tried ChatGPT, seven have only tested it, five use it sometimes and only two use it regularly.

4.2   CT and Rule-Based Programming

As students had already received formal education in basic programming according to the Swedish curriculum and as confirmed by the results of the pre-intervention questionnaire, we decided to test their skills at an early stage of the workshop series. Session 2 was entirely focused on assessing their skills in CT and basic rule-based programming (see Table 1). This session used challenge-based activities where students had to program wheeled robots equipped with different types of sensors. The students were given six challenges of increasing difficulty (challenge 1 was the easiest and challenge 6 was the hardest) requiring the use of different CT concepts such as loops, and conditionals. Figure 1 shows the level of completeness that the students achieved for each of the six challenges. These challenges ranged from constructing simple and short algorithms to more complex ones that would require the use of specific programming concepts like working with different data types and using logic/arithmetic operators and variables. Also, these challenges required students to decide the type of sensors that could work best for the specific challenge they were working on. For this activity, students worked in pairs.

	Challenge 1	Challenge 2	Challenge 3	Challenge 4	Challenge 5	Challenge 6
Group 1	Complete	Complete	Complete	Complete	Complete	Complete
Group 2	Complete	Complete	Complete	Incomplete	Complete	Incomplete
Group 3	Complete	Complete	Incomplete	Not done	Incomplete	Not done
Group 4	Complete	Incomplete	Incomplete	Not done	Not done	Not done
Group 5	Complete	Incomplete	Not done	Not done	Not done	Not done
Group 6	Complete	Incomplete	Not done	Not done	Not done	Not done
Group 7	Complete	Complete	Not done	Not done	Not done	Not done
Group 8	Complete	Incomplete	Not done	Not done	Not done	Not done
	Completed: 100%	Completed: 50%	Completed: 25%	Completed: 12.5%	Completed: 25%	Completed: 12.5%

Fig. 1. Level of completion of rule-based programming activities in session 2

Figure 1 shows that the completion level decreased considerably as the challenges increased in difficulty and required more complex CT concepts. Only the first challenge (which required the construction of a short algorithm based on a single loop) was completed by all groups. However, challenges 4, 5, and 6 (which required using conditionals, logic/arithmetic operators, and variables) were completed successfully by only one group. The relatively low performance of the students during these challenged-based programming activities suggests that they have difficulty applying programming concepts and expressing abstract ideas through algorithmic thinking. In addition, difficulties in debugging code might also have contributed to the low completion rates in many of the challenges.

4.3 AI and ML

We covered AI-related topics during the workshop sessions 1, 3, 4, and 5 (see Table 1). We attempted to gain insights on the conceptualizations that students had regarding AI technology. In addition, we introduced the students to some fundamental concepts of AI such as the role of datasets and training ML models.

Students' Conceptualizations of AI. At the beginning of session 1, we used a real time data collection tool to get a picture of some students' notions and preconceptions regarding AI. Most students associate AI with robots and ChatGPT. Out of the 16 students that answered the multiple-choice questions 11 consider AI to be intelligent, nine that it can think and understand but only three think it has emotions. Students' main worries about AI are that it might replace humans, spread misinformation, take over the world, and that it could be detrimental to the human ability to think by themselves. The positive attitudes to AI were related to efficiency and automation of tasks while only a few considered it positive for creativity. Surprisingly, since most consider AI to be intelligent, they almost unanimously express that a goldfish is more intelligent than systems like ChatGPT or autonomous vehicles despite all these systems using AI.

Image Classification - Google Teachable Machine. During session 1, (see Table 1), students worked in groups of two to three, they were given an imbalanced dataset with two classes, dogs and cats. The data was divided into a training set and a test set (Payne 2019). After training the classifier with the training data the students used the test data to see the accuracy with which the black boxed model could predict the correct class. The students then wrote down why they thought that the model had better accuracy predicting one class than the other. The more obvious reasons such as one dataset being overrepresented in numbers were quickly mentioned. The students then started looking at the variety of dogs and cats represented in these datasets. One class was more homogenous than the other and then these features were discussed in detail. Students were tasked to retrain their model and had to go looking for images available to create representational and fair datasets. The discussion after the activity suggests that students could relate this knowledge about data representation to other contexts and implications thereof such as unfair and less accurate predictions for underrepresented groups.

Learning the Fundamentals of Machine Learning. In session 3 (see Table 1) students were involved in activities about supervised and unsupervised machine learning using their own data. We prepared some Python code using machine learning and data visualisation libraries. Students had to note down their shoe size and heights on a post-it note, and all the values were entered into a data frame with labels and one data frame unlabelled. Running the script produced visualisations of the data using linear regression to classify the data. The data was inspected and although high accuracy in predicting the students' gender was achieved some students were misclassified. The same situation happened when using K means clustering on the unlabeled data. Students expressed two main insights from this exercise, 1) misclassification due to under/over representation of certain (normal) values, and 2) how seemingly irrelevant data such as height and shoe size can predict information that might be considered sensitive, such as gender.

4.4 Relationship Between AI and CT

To get a deeper insight on students' understanding of AI and ML concepts, we engaged them in some exercises and hands-on activities where they had to reflect on the relationship between AI-based methods and traditional rule-based programming.

Conceptualising the Relationship Between CT and AI. To this end, the students were given a written exercise where they were asked to explain the main characteristics of AI and rule-based programming, and to find possible differences and/or similarities between these two approaches. We found that the main descriptions mentioned by the students about AI can be connected to some of the main AI competencies listed in the framework for AI literacy from Long and Magerko (see table in Figshare for more details). Among the most frequent characteristics of AI given by students are the ability to make decisions and to learn from examples. Furthermore, students mention other attributes of AI such as being able to figure out things, draw conclusions, work for a wide range of situations, and that it can answer questions. In this exercise, 11 out of 12 students managed to provide a definition of AI and ML that is at least partly related to some of the main AI competencies listed in the framework for AI literacy.

As for the descriptions regarding rule-based programming, most students mentioned that this method is based on the construction of programs containing rules. Although they expressed this idea with slight differences from one another, considering the CT concepts from the framework by Brennan and Resnick, we find that 11 out of 12 students connected rule-based programming with the CT concept of sequence (see table in Figshare for more details). The students referred to the sequential nature of rule-based programming emphasizing aspects such as the fact that it is a set of rules, a code containing instructions that a computer follows, and that it is a step-by-step process. The students even mentioned other aspects attributed to rule-based programming, which go beyond the CT framework. For instance, some students mentioned that systems built on rule-based programming cannot learn, that they are suitable for simple tasks only, that they cannot deal with situations that go beyond the rules defined in their original programs and that a person is needed for programming the rules. Although it is debatable how accurate these definitions are, they are a reminder that apart from the technical concepts that students managed to understand during the workshops regarding the main characteristics of ML and rule-based programming, there also might be some notions that are at least partly based on personal or collective perceptions and preconceptions.

Practically Integrating CT and AI. The last challenge-based exercise in the workshop series (see session 4 in Table 1) was a hands-on activity aimed at testing students' abilities to integrate rule-based programming with AI techniques contributing to a working application. The main task was that students working in pairs had to create and train ML models (using Machine Learning for Kids) and then integrate them into a script built in Scratch. For this activity students would choose their challenges based on their skills and interests. The results from this activity show that in general, students did not encounter major problems when training the ML models (only 2 out of 9 groups could not complete the entire training process). Taking into consideration that students had already been working on training other ML-based systems during the workshop series (e.g. when they worked training the Google Teachable Machine or when training Misty the robot to recognize faces), this could account for the higher success rate in this phase of the exercise. Also, students showed a higher level of engagement when training ML models as they found it a fun task. The situation changed considerably during the second phase of this exercise where students had to integrate the trained models in a script built in Scratch, where only 2 out of 9 groups managed to carry out this task successfully.

The students encountered complications using CT and programming concepts, such as conditionals and logic operators when constructing the algorithms. These difficulties that students experienced when programming in Scratch were similar to those they had when programming the wheeled robots during session 2. This suggests that, although students had previous experience doing programming, they still struggle to understand some CT and programming concepts and thus, it is difficult for them to abstract an idea for a solution to a given problem and express it using algorithmic logic.

5   Discussions

Introducing core concepts related to AI literacy in K-12 education poses several challenges. Our study highlights some of these challenges and in this section, we discuss potential ways to approach these issues.

Conceptualisations of AI. Students hold beliefs of what AI is and what it is capable of. Some of the pre-existing beliefs persist even after students have engaged in workshops related to data and ML. Although students managed to understand basic concepts of AI in the context they were taught during the workshops, transferring this specific knowledge to other different contexts proved challenging, which could explain why some preconceptions about AI persist. Furthermore, even if our results suggest that students acquired a certain understanding about AI and ML methods, this knowledge is basic and superficial and therefore it is hard to assess to what extent it derives from the teaching interventions rather than from other sources more related to social preconceptions of this technology. These possible preconceptions may pose some difficulties especially when teaching AI-related topics to students that have not yet received formal education in this subject.

Limited Knowledge in Programming. Based on current Swedish curriculum, programming is part of mathematics and technology from first class to ninth grade, however our results show that students' actual knowledge in programming is limited. One issue might be that there are no clear criteria for assessing programming abilities and there are no specific learning objectives either. It is then up to the teacher to integrate programming in their own particular manner. Another possible issue is that programming is only taught in the context of mathematics and technology aiming at doing maths and solving technological problems. However, this poses a big limitation as in the context of mathematics the use of programming might tend to be isolated from the larger socio technical perspectives that exist in the real world which may result in lack of interest and engagement from the students to learn traditional rule-based programming. This situation is in stark contrast to the interdisciplinary nature of AI where students can use this technology to engage with other subjects such as language, art, or biology. Our results reflect this as students showed higher levels of engagement when working with systems based on AI techniques.

The Way Towards Adequate Digital Competences. Being digitally competent according to the Swedish Agency for Education requires abilities to 1) solve problems and put ideas into action 2) use and understand digital tools and media, 3) have a critical and responsible approach, and 4) understand societal implications of digitalisation. The ongoing process of digitalization in education will require study plans to include topics regarding the use of emergent technologies like AI and ML where students can explore the possibilities offered by these new technologies to devise and evaluate solutions in modern society. However, it is precisely in this area where we identified some of the main challenges when introducing middle school students to AI, as they showed clear difficulties when engaging in activities where they were required to devise a solution solving different problems and tasks. Students seem to have problems applying CT and programming concepts when devising solutions. It is therefore necessary that students understand the importance of CT and AI as both methods are needed when developing advanced solutions to address complex problems. The interrelation between traditional rule-based programming and AI are important aspects for fostering AI literacy and digital competences among K-12 students. Thus, understanding the fundamentals of CT and traditional programming is a prerequisite for reaching a deeper comprehension of what AI technology is, what it can do and how it can help us develop our problem-solving skills.

Contextualising ML and AI When Opening the Black Box. Our results also indicate that engaging with ML and AI on abstract levels and removing the complexity from the process gave students the impression that AI and ML are easier to implement than CT concepts and traditional programming. Therefore, working with the outcomes of black boxed ML models in conjunction with visualizing ML concepts such as linear regression and clustering (see session 3 in Table 1) seems important to facilitate a deeper understanding of the complexity of ML and AI. This method could help demystify AI as the so-called Eliza effect as well as the tale-spin effect (Long & Magerko 2020) are persistently present in students' responses to how they conceptualize and perceive AI.

6   Conclusions, Limitations and Future Work

Our study identified some potential challenges when introducing AI in elementary school education. There are a series of preconception that students have about what AI technology is and what it can do. After conducting the hands-on workshops, the students managed to acquire some basic understanding of AI and ML. However, this knowledge may still contain notions stemming from individual or collective preconceptions, which may constitute a hinder to a deeper comprehension of this subject. Traditional rule-based programming is a relevant prerequisite to reach a higher level of digital competence and AI literacy. Nevertheless, students exhibited limited skills in programming even if they have already received education in this topic in school. An integrated and complementary teaching of CT and AI concepts is necessary to foster AI literacy and problem-solving skills among students.

This empirical study had a limited number of participants who all attended the same school. Therefore, our results are more exploratory than generalizable. Nevertheless, they provide insights regarding the challenges of introducing a new topic of study in K-12. Due to GDPR regulations for data and privacy protection, we took no photos during this study. In our upcoming endeavors, we aim to further refine and implement the workshop concepts and content outlined in this paper across a broader spectrum of schools. This expansion will allow us to glean additional insights and gather more comprehensive data to advance our explorations.

References

Bocconi, S., et al.: Exploring the field of computational thinking as a 21st century skill. In: EDULEARN16 Proceedings, pp. 4725–4733, IATED (2016)

Brennan, K., Resnick, M.: New frameworks for studying and assessing the development of computational thinking. In: Proceedings of the 2012 Annual Meeting of the American Educational Research Association, vol. 1, p. 25. Vancouver, Canada (2012)

Conde, M.Á., et al.: RoboSTEAM-a Challenge based learning approach for integrating STEAM and develop computational thinking. In: Proceedings of the Seventh International Conference on Technological Ecosystems for Enhancing Multiculturality, pp. 24–30 (2019)

Dohn, N.B., Kafai, Y., Mørch, A., Ragni, M.: Survey: artificial Intelligence, computational thinking and learning. KI-Künstliche Intelligenz 36(1), 5–16 (2022)

Elo, S., Kyngäs, H.: The qualitative content analysis process. J. Adv. Nurs. 62(1):107–15 (2008)

Giroux, H.A.: Education incorporated? Educ. Leadersh. 56(2), 12–17 (1998)

Grover, S., Pea, R.: Computational thinking in K–12: a review of the state of the field. Educ. Res. 42(1), 38–43 (2013)

Grover, S., Pea, R.: Computational thinking: a competency whose time has come. Comput. Sci. Educ. Perspect. Teach. Learn. School 19(1), 19–38 (2018)

Knobel, M., Lankshear, C.: Studying new literacies. J. Adolescent Adult Literacy 58(2), 97–101 (2014)

Heintz, F.: The computational thinking and artificial intelligence duality, Computational Thinking Education in K-12: artificial intelligence literacy and physical computing, p. 143. The MIT Press (2022)

Kafai, Y.B., Burke, Q.: Computational participation: Teaching kids to create and connect through code. Emerging research, practice, and policy on computational thinking, pp. 393–405 (2017)

Long, D., Magerko, B.: What is AI literacy? Competencies and design considerations. In: Proceedings of the 2020 CHI Conference on Human Factors in Computing Systems, pp. 1–16 (2020)

Merchant, G.: Reading with technology: the new normal. Education 3–13 49(1), 96–106 (2021) https://doi.org/10.1080/03004279.2020.1824705

Miao, F., Holmes, W., Huang, R., Zhang, H.: AI and education: a guidance for policymakers. UNESCO (2021)

Payne, B.H.: An ethics of artificial intelligence curriculum for middle school students. MIT Media Lab Personal Robots Group (2019)

Pedaste M, et al.: Phases of inquiry-based learning: Definitions and the inquiry cycle. Educ. Res. Rev. 14, 47–61 (2015)

Selby, C., Woollard, J.: Computational thinking: the developing definition (2013)

Skolverket. Curriculum for the compulsory school, preschool class and school-age educare (2018). Retrieved March 2024. https://www.skolverket.se/download/18.31c292d516e7445866a218f/1576654682907/pdf3984.pdf

Skolverket [the Swedish National Agency for Education] (2022). [Four aspects of digital competence]. Retrieved online at (2022) https://www.skolverket.se/om-oss/var-verksamhet/skolverketsprioriterade-omraden/digitalisering/fyra-aspekter-av-digital-kompetens. Accessed 20 Oct 2022

Tedre, M., Denning, P., Toivonen, T.: CT 2.0. In: Proceedings of the 21st Koli Calling International Conference on Computing Education Research, pp. 1–8 (2021)

Touretzky, D., Gardner-McCune, C., Martin, F., Seehorn, D.: Envisioning AI for K-12: what should every child know about AI?. In: Proceedings of the AAAI Conference on Artificial Intelligence, vol. 33, no. 01, pp. 9795–9799 (2019)

Velander, J., Taiye, M.A., Otero, N., Milrad, M.: Artificial Intelligence in K-12 Education: eliciting and reflecting on Swedish teachers' understanding of AI and its implications for teaching & learning. Educ. Inform. Technol. (2023)

Velander, J., Otero, N., Milrad, M.: What is Critical (about) AI literacy? exploring conceptualizations present in AI literacy discourse. exploring conceptualizations present in AI literacy discourse. Springer Nature Switzerland AG. In Framing Futures in Postdigital Education: Critical Concepts for Data-driven Practices (in press)

Vinnervik, P., Bungum, B.: Programming in the curriculum for compulsory school: how is it represented in Nordic countries. In: Science Education in the light of Global Sustainable Development-trends and possibilities: Proceedings of the 13th Nordic Research Symposium on Science Education, pp. 193–201 (2021)

Vuorikari Rina, R., Kluzer, S., Punie, Y.: DigComp 2.2: The Digital Competence Framework for Citizens-With new examples of knowledge, skills and attitudes. Joint Research Centre, Seville site (2022)

Wolff, A., Gooch, D., Montaner, J.J., Rashid, U., Kortuem, G.: Creating an understanding of data literacy for a data-driven society. J. Commun. Inform. (2016)

Zerega, R., Milrad, M.: Computational thinking & artificial intelligence in K-12 education: two distinct but still complementary worlds. In: Milrad, M., et al. (eds.) Methodologies and Intelligent Systems for Technology Enhanced Learning, 13th International Conference. MIS4TEL 2023. LNNC, vol. 764. Springer, Cham (2023). https://doi.org/10.1007/978-3-031-41226-4_22

The Learning Effects in Immersive Technologies

Kai Pata[✉] and Terje Väljataga

Tallinn University, Narva Road 25, 10120 Tallinn, Estonia
`kai.pata@tlu.ee`

Abstract. In this comparative literature study we collected a sample of recent (from the period of 2020–2022) empirical and meta-studies of immersive technologies in practice based learning. The study explored empirical and meta-studies of the virtual learning environments, extended and augmented reality environments regarding which learning outcomes were measured in cognitive, metacognitive, affective and psychomotor, behavioural and embodied learning domains. It was found that although there are plenty of experiments with immersive technologies, there is not sufficient clarity on what way these technologies may support practice-based learning. We found that the learning experiments with immersive technologies lack the collaborative coworking dimensions, and the learning process results are conceptualised at individual learner level. Research in empirical studies is focusing only on limited types of learning outcomes. The data analysis focused on learning effects and obstacles in practice-based learning with immersive technologies and synthesised the design principles that could support immersive learning.

Keywords: Immersive technologies · Learning effects · Learning design

1 Introduction

The opportunities of immersive learning technologies in changing practice-based learning have been manifested. In this article, we conceptualise practice-based learning as a comprehensive term for instructional approaches that encourage learners to actively engage in the learning process (e.g. inquiry learning, action learning, experiential learning, simulations, gamified learning). We view in this paper the immersive learning technologies: virtual reality (VR), augmented reality (AR), mixed reality (MR) and extended reality (XR) as learning environments for practice-based learning. There is not a clear empirical based overview, what learning effects have been explored in practice-based learning with immersive technologies. To discover what learning effects have been studied in practice-based learning with immersive technologies the literature analysis was conducted. The data analysis focused on the following research questions: Which learning effects and obstacles do immersive technologies create for learners in practice-based learning? What design principles could support immersive learning?

T. E. Kim et al. (Eds.): MIS4TEL 2024, LNNS 1274, pp. 98–108, 2025.
https://doi.org/10.1007/978-3-031-84170-5_9

2 Methods

The sample of research papers of immersive learning technologies was driven from SCO-PUS database using the keywords of VR, AR, XR, using multimedia new approaches, and constraint period of 2020–2022. The sample of learning effects papers was formed of 226 papers, from which 65 papers were selected for further analysis. For exploring the possible learning effects of immersive learning technologies on users we selected four domains: cognitive effects, metacognitive effects, affective effects, and psychomotor, behavioural and embodied [1] effects.

The analytical process was structured with the questions: Type of technologies and media used in experiments; Type of learning effects (Cognitive, metacognitive, affective, psychomotor, behavioural and embodied); Description of the learning obstacles. The literature analysis was done by three researchers. The coding process was tested for interrater reliability with some sample articles. The data analysis was done using the sorting method (the results are mostly presented in tables) and qualitative synthesis of the findings.

3 Results

3.1 Cognitive Learning Effects with Immersive Technologies

In most of the research papers the conceptual background from the cognitive information processing paradigm was used [2], and the social constructivist [3], situated and distributed cognition [4, 5] and embodied cognition aspects [6–8] were addressed seldom. We did not find any studies exploring empirically beyond the individual level of cognitive learning - no group cognition aspects were explored such as conceptual shared understandings, coherence etc.

The virtual environments were praised for increased object visualisation opportunities [9] and presenting non-existent things [10]. Students' understandings were found to be improved [11–13], that could be influenced by more realistic, authentic and interdisciplinary as well as with greater complexity situations in immersive and gamified environments using narratives in serious games [14–20], viewing body language [21], picking up objects and examining these [13, 22], the interaction with people and group work in the virtual environment [12], experiencing VR time and space [23]. Interaction with people in immersive environments gave a deeper impression of the learning content [12].

Regarding negative aspects, first being anxious of losing face in front of their classmates when asked to perform while the whole class was present [24], social comparison pressure and interpersonal communication difficulties [25], asymmetric team interactions [26], interacting with avatars without facial expressions replacing students [27], possible cybersecurity threats were negative social presence and privacy related aspects in virtual environments [28].

Furthermore, immersive learning technologies were found to cause cognitive load issues in the working memory information processing [13, 17]. Evidence of several cognitive learning issues were not paying attention to learning objectives in virtual reality [29], being too much absorbed in the simulations [30], feeling bored in the virtual reality that is frequently used, or having fatigue [27, 31], and having integration difficulties [17]. Visual and sound level challenges, and reading speed issues were faced by some students who had specific impairments. The lessened opportunity to talk or move and operate bodily was indicated in virtual environments [14, 32, 33]. This may have a negative effect on cognitive learning such as auditory information retention [13]. Also there were issues in interpreting micro-gestures in virtual environments [28]. We found no studies about developing misconceptions in the interactive media environments that would relate with the level of authenticity instead of misleading modelling. One study indicated that abstract conceptual understanding could be improved [13]. In another study the issues of signifying objects in virtual reality was highlighted that caused difficulties to remember semantic information [30].

3.2 Metacognitive Learning Effects with Immersive Technologies

Metacognitive learning effects [34] such as self-regulation [35], self-directed learning [36] were less frequently studied in the sample of observed papers. To support metacognitive learning with immersive technologies the learner autonomy [37, 38] was highlighted. The instructional design elements such as experiencing different role perspectives [37], debriefing for developing insights [33], self-reflection and feedback from teachers [39], peers or virtual characters [20, 40] were considered important to understand actions in virtual reality. In addition, monitoring of students' interactions with AI that prompted students to give feedback was considered an important instructional design element [40].

Self-efficacy was one of the frequently measured constructs related with metacognition and affections [16, 17, 27, 41–43]. The development of self-regulation skills [44, 45] with the scaffolding prompts [27] was associated with better attention, awareness and cognitive practice in virtual environments [44]. Also practising specific dialogic interactions triggered by scaffolding prompts provided by chatbots [39, 47] or characters in the virtual world [48] were found to improve students' communicative abilities and develop identity [48]. Embodying a character in virtual reality, interactivity and game elements made participants more involved, gave them a feeling of control over the course of the narrative [37] and increased learners' confidence [16] and agency, which may lead to improved learning outcomes [49].

However, in addition to positive aspects, the virtual environments were sometimes associated with negative emotions and lower perceived self-control [46].

3.3 Affective Learning Effects with Immersive Technologies

Affective learning in immersive environments was most often explored through motivational aspects [50, 51]. In our sample of research papers we found that realistic elements in virtual reality environments [21, 28], self-location within the story environment [52], affective scaffolding with AI chatbots [39], social interaction in virtual environments [53] co-presence among users [52], gamification [15, 54], and formative assessment in virtual environments [9] were increasing learners' motivation [42] and in some cases also cognitive processing and performance [52, 55].

Negative effects of cognitive learning in immersive environments were of the affective kind: emotions like fear [56]. The negative emotions were decreased by virtual world anonymity [17]; attentive and curious AI chatbots telling jokes and fun facts [39]; fun feeling from gamification where to explore different options autonomously [20, 57], collectivistic versus individualistic user orientations that moderate the effects of value on attitudes in a gamification context [58], the avatars that help to allay worries about being judged negatively [59], open communication processes [43], small group learning immersive experience [17, 47] and interactive switching of scenarios improving learners' autonomy, active engagement, and collaboration with partners [17]. It was reported that the participants feel less nervous in the virtual classroom [22].

3.4 Behavioural, Psychomotor and Embodied Cognition Effects with Immersive Technologies

Psychomotor [60, 61] and embodied learning [62] effects were the least studied in immersive learning technologies because the sensorimotor contingencies such as immersion, spatial and virtual presence, spatial location, identity have not been strongly related with the traditional cognitive and metacognitive, affective and psychomotor learning effects that are currently mainly considered as more important learning outcomes. There are not enough studies about how multisensory inputs and information processing takes place. These sensorimotor contingencies of the body are not yet well understood, but it is assumed that cognitive and physical involvement is interrelated in virtual environments [47]. The virtual reality environment, or augmented reality is formulated as a sensory-motor contact with the world, with the organ serving as the mediator in the process. It is the sensation and vision organ and the kinesthetic structure that constructs knowledge and allows for complete body interaction, allowing users to visualise the world by perceptual learning [39]. VR is supposed to activate the brain to support a user's natural inclination to engage sensorimotor contingencies [28]. There are some design suggestions that narratives, authenticity realism and interaction with the virtual environment and augmented reality might be improving the sensorimotor contingencies [21, 63, 64], the self-evaluation of performance [63]. The egocentric view of the user [10] in the centre of the space is also believed to be developing these sensorimotor contingencies and might improve the skill transition. Psychomotor and embodied learning effects with immersive technologies relate with some negative physical discomforts that some learners perceive - feeling tired or physical discomfort [17, 65], motion sickness and nausea [43, 66], dizziness [67, 68], claustrophobia, migraine [13, 17, 21, 23, 29, 43, 69, 70, 71] or identity or reality confusion [67]. Simulated movement and real walking is suggested as a remedy instead of flying in virtual spaces [71, 72], also it is important to have human bodily features like the number of limbs, size [67], the virtual world should not outsize the available physical space [33]. Sensorimotor and behavioural learning also causes some accessibility problems for people with visual, motor, hearing and cognitive impairments.

4 Design Principles for Immersive Learning Environments

It was evident from the papers that in immersive environments the cognitive, metacognitive, psychomotor and affective learning design elements and learning effects are interwoven. The learning experiences with immersive technologies point out the important aspects of learning design for meaningful and safe learning episodes with immersive technologies, many of them not present in traditional modes of learning and teaching. Below we summarise based on the findings of our study a set of potential design principles that have had shown effects on learning.

Cognitive and Psychomotor Learning

Develop authentic situations with immersive storytelling [73] for better understanding [11, 14, 23], stimulation of imagination [10], improvement of knowledge and skill transfer with multisensory opportunities that can help to relate interdisciplinary domains and understand the complexity [16–18], but consider that they may also cause integration difficulties [17].

Consider that the sense of being immersed in an unmediated reality may impact negatively on cognitive load [11, 13, 17, 74] and concentration on task [29, 30, 75]. Consider graphic quality - object realism level [23], image quality [14, 17, 19], rendering speed, visibility issues [37, 75], noise and sound level, speech quality [76]. Use the surrounding environment to anchor and help encode learning [11]. Embed meaningful textual information but be cautious to create need to scroll, press buttons etc. [53].

Consider that the reading speed may be lower in immersive reality [17], learning from auditory information may be lower than learning from visual information [13], and learning specific facts, names, dates may be hindered [30].

Consider learners' sensory [19, 43], cognitive and motor impairments (e.g. BCI-illiteracy) and negative psycho-physical effects like fear at heights, anxiety, cybersickness, stress, risk-taking, headache, dizziness, nausea, disorientation, blurred and double vision, eyestrain, focusing difficulty, heart rate, difficulty at breathing/speaking [13, 17, 37, 43, 56, 68, 70, 71] and learning distractions [44, 75]. Provide a physically comfortable [12, 17, 29], adjusted to persons' eyes [21] and a safe environment. The position of the learner in situations could be central, being the main character [52] as having an egocentric point of view may improve skills transfer to real world situations [10].

Use actual rather than teleported movement [77] and see that the virtual space would not outsize available physical space [33].

Give to the learners control over their body and movement [14, 78] and allow learners to use their body, mediated limbs with controllers and skin sensors to promote an understanding of abstract concepts [13, 27, 31, 56].

Make use of the repeated practice [27].

Activate learners' prior knowledge [42].

Diversify the learners' exposure to practical knowledge to improve recall [9, 31, 42].

Metacognitive Learning

Promote learner agency, feeling of control [37] with active exploration [64] and making their own choices [20].

Facilitate step-by-step procedural learning [45], reflection on action [44] and self-evaluation [63].

Create an assessment integrated into teaching [9].

Affective Learning

Provide in an immersive space motivation management to hold the interest with gamification elements, avatars, in-depth details [37], perceived authenticity [12], interactivity that involves personally [52], and develop positive emotions - fun, enjoyment [79], happiness, excitement [47].

Maintain negative emotions of boredom [27], fatigue [31, 65] social nervousness [17], fear and anxiety [24].

Negative scenarios may relate with lower perceived control [46].

Social Learning

Provide an informal learning environment with the atmosphere of mutual trust and empathy [23].

Provide interaction with learners or avatars to improve understanding [12, 27, 47] or add the narrators [64].

Virtual characters' body language may help to guess the meaning [21].

Scaffold the learners [20, 22].

Provide feedback to improve students' learning effect [33].

Consider the privacy of learners and cybersecurity [28].

In conclusion, several negative and challenging aspects for learning emerged with immersive technologies can be reduced or eliminated with careful consideration of learning design. The presented learning design aspects are theoretical considerations that can be taken into account for further learning designs with immersive technologies.

5   Conclusion

The immersive technologies create the medium and object for practice where cognitive, metacognitive, affective and psychomotor and embodied learning effects may be differently experienced and facilitated. Yet, our literature analysis indicates that theoretically explaining the immersive learning environments through learning are not well studied regarding different phenomena and concepts. We found that the learning experiments with immersive technologies lack the collaborative coworking dimensions, and the learning process results are conceptualised at individual learner level. Research in empirical studies is focusing only on limited types of learning outcomes. There is still a lot of technology optimism without strong evidence-based knowledge of how immersive technology affects learning.

Acknowledgments. Research supported by the e-DIPLOMA, project number 101061424, funded by the European Union. Views and opinions expressed are, however, those of the authors only and do not necessarily reflect those of the European Union or the European Research Executive Agency (REA). Neither the European Union nor the granting authority can be held responsible for them.

References

1. Bloom, B.S., Engelhart, M.D., Furst, E.J., Hill, W.H., Krathwohl, D.R.: Taxonomy of educational objectives: the classification of educational goals—Handbook I, cognitive domain. David McKay, New York, NY (1956)
2. Atkinson, R.C., Shiffrin, R.M.: Human memory: a proposed system and its control processes. Psychol. Learn. Motiv. **2**, 89–195 (1968)
3. Lave, J., Wenger, E.: Situated learning: legitimate peripheral participation. Cambridge university press (1991)
4. Brown, J.S., Collins, A., Duguid, P.: Situated cognition and the culture of learning **18**(1), 32–42 (1989)
5. Hutchins, E.: Cognition in the wild. MIT press (1995)
6. Gibson, E.J.: Exploratory behaviour in the development of perceiving, acting, and the acquiring of knowledge. Annu. Rev. Psychol. **39**(1), 1–42 (1988)
7. Lakoff, G., Johnson, M.: Philosophy in the flesh: the embodied mind and its challenge to western thought. Basic Books (1999)
8. Shapiro, L.A.: Embodied cognition: lessons from linguistic determinism. Philos. Top. **39**(1), 121–140 (2011)
9. Zhang, H., et al.: Hotspots and trends of virtual reality, augmented reality and mixed reality in the education field. In: 2020 6th International Conference of the Immersive Learning Research Network (iLRN), pp. 215–219. IEEE (2020)
10. Barrett, A.J., Pack, A., Quaid, E.D.: Understanding learners' acceptance of high-immersion virtual reality systems: Insights from confirmatory and exploratory PLS-SEM analyses. Comput. Educ. **169** (2021)
11. Baceviciute, S., Terkildsen, T., Makransky, G.: Remediating learning from non-immersive to immersive media: using EEG to investigate the effects of environmental embeddedness on reading in Virtual Reality. Comput. Educ. **164**, 104122 (2021)
12. Yang, F., Goh, Y.M.: VR and MR technology for safety management education: An authentic learning approach. Saf. Sci. **148**, 105645 (2022)
13. Di Natale, A.F., Repetto, C., Riva, G., Villani, D.: Immersive virtual reality in K-12 and higher education: a 10-year systematic review of empirical research. Br. J. Edu. Technol. **51**(6), 2006–2033 (2020)
14. Eiris, R., Jain, A., Gheisari, M., Wehle, A.: Safety immersive storytelling using narrated 360-degree panoramas: a fall hazard training within the electrical trade context. Saf. Sci. **127**, 104703 (2020)
15. Dehghanzadeh, H., Fardanesh, H., Hatami, J., Talaee, E., Noroozi, O.: Using gamification to support learning English as a second language: a systematic review. Comput. Assist. Lang. Learn. **34**(7), 934–957 (2021)
16. Erdogmus, E., Ryherd, E., Diefes-Dux, H.A., Armwood-Gordon, C.: Use of virtual reality to improve engagement and self-efficacy in architectural engineering disciplines. In: 2021 IEEE Frontiers in Education Conference (FIE), pp. 1–7. IEEE (2021)
17. Bahari, A.: Affordances and challenges of teaching language skills by virtual reality: aA systematic review (2010–2020). E-Learn. Digital Media **19**(2), 163–188 (2022)
18. Cooper, V.A., Forino, G., Kanjanabootra, S., von Meding, J.: Leveraging the community of inquiry framework to support web-based simulations in disaster studies. Internet High. Educ. **47**, 100757 (2020)
19. Galeote, F.D., Hamari, J.: Game-based climate change engagement: analysing the potential of entertainment and serious games. In: Proceedings of the ACM on Human-Computer Interaction, vol. 5(CHI PLAY), pp. 1–21 (2021)

20. Buijs-Spanjers, K.R., Harmsen, A., Hegge, H.H. et al.: The influence of a serious game's narrative on students' attitudes and learning experiences regarding delirium: an interview study. BMC Med. Educ. **20**(289) (2020)
21. Akgün, M., Atici, B.: The effects of immersive virtual reality environments on students' academic achievement: a meta-analytical and meta-thematic study. Participatory Educ. Res. **9**(3), 111–131 (2022)
22. Chen, C.Y., Chang, S.C., Hwang, G.J., Zou, D.: Facilitating EFL learners' active behaviours in speaking: a progressive question prompt-based peer-tutoring approach with VR contexts. Interact. Learn. Environ. 1–20 (2021)
23. Li, P., Fang, Z., Jiang, T.: Research into improved distance learning using VR technology. Front. Educ. **7** (2022)
24. Yang, Q.F., Chang, S.C., Hwang, G.J., Zou, D.: Balancing cognitive complexity and gaming level: effects of a cognitive complexity-based competition game on EFL students' English vocabulary learning performance, anxiety and behaviours. Comput. Educ. **148**, 103808 (2020)
25. Chan, C-S. Chan, Y-H., T Agnes Fong, T.H.: Game-based e-learning for urban tourism education through an online scenario game. Int. Res. Geograph. Environ. Educ. **29**(4), 283–300 (2020)
26. Divekar, R.R., et al.: Foreign language acquisition via artificial intelligence and extended reality: design and evaluation. Comput. Assist. Lang. Learn. 1–29 (2021)
27. Chen, C.-J.: Immersive virtual reality to train preservice teachers in managing students' challenging behaviours: a pilot study. Br. J. Edu. Technol. **53**, 998–1024 (2021)
28. Dwivedi, Y. =K., et al.: Metaverse beyond the hype: Multidisciplinary perspectives on emerging challenges, opportunities, and agenda for research, practice and policy. Int. J. Inform. Manage. **66**, 102542 (2022)
29. Arayaphan, W., Sirasakmol, O., Nadee, W., Puritat, K.: Enhancing intrinsic motivation of librarian students using Virtual reality for education in the context of culture heritage museums. TEM J. **11**(2) (2022)
30. Ebadi, S., Ebadijalal, M.: The effect of Google Expeditions virtual reality on EFL learners' willingness to communicate and oral proficiency. Comput. Assist. Lang. Learn. **35**(8), 1975–2000 (2022)
31. Johnson-Glenberg, M.C., Bartolomea, H., Kalina, E.: Platform is not destiny: embodied learning effects comparing 2D desktop to 3D virtual reality STEM experiences. J. Comput. Assist. Learn. **37**(5), 1263–1284 (2021)
32. MacWhinney, B.: A shared platform for studying second language acquisition. Lang. Learn. **67**(S1), 254–275 (2017)
33. Clack, L., Hirt, C., Kunz, A., Sax, H.: Experiential training of hand hygiene using virtual reality. Recent Advances in Technologies for Inclusive Well-Being: Virtual Patients, Gamification and Simulation, pp. 31–42 (2021)
34. McComas, W.F.: Metacognition. In: McComas, W.F. (ed.) The Language of Science Education, pp. 63–63. SensePublishers, Rotterdam (2014). https://doi.org/10.1007/978-94-6209-497-0_55
35. Zimmerman, B.J., Schunk, D.H.: Self-regulated learning and academic achievement: theoretical perspectives, 2nd edn. Lawrence Erlbaum Associates Publishers (2001)
36. Knowles, M.S.: Self-directed learning: a guide for learners and teachers (1975)
37. DeWitt, D., Chan, S.F., Loban, R.: Virtual reality for developing intercultural communication competence in Mandarin as a Foreign language. Educ. Tech. Res. Dev. **70**(2), 615–638 (2022)
38. Li, M., Chau, P.Y., Ge, L.: Meaningful gamification for psychological empowerment: exploring user affective experience mirroring in a psychological self-help system. Internet Res. **31**(1), 11–58 (2021)
39. Kuhail, M.A., Alturki, N., Alramlawi, S., Alhejori, K.: Interacting with educational chatbots: a systematic review. Educ. Inf. Technol. **28**(1), 973–1018 (2022)

40. Villegas-Ch, W., Arias-Navarrete, A., Palacios-Pacheco, X.: Proposal of an architecture for the integration of a Chatbot with artificial intelligence in a smart campus for the improvement of learning. Sustainability (Switzerland) **12**(4) (2020)

41. Azila-Gbettor, E.M., Mensah, C., Abiemo, M., Bokor. M.: Predicting student engagement from self-efficacy and autonomous motivation: a cross-sectional study. Cogent Educ. **8**(1), 1942638 (2021)

42. Asad, M.M., Naz, A., Churi, P., Tahanzadeh, M.M.: Virtual reality as a pedagogical tool to enhance experiential learning: a systematic literature review. Educ. Res. Int. 1–17 (2021)

43. Elzie, C.A., Shaia, J.: A pilot study of the impact of virtually embodying a patient with a terminal illness. Med. Sci. Educ. **31**, 665–675 (2021)

44. Drigas, A., Mitsea, E., Skianis, C.: Virtual reality and metacognition training techniques for learning disabilities. Sustainability **14**(16), 10170 (2022)

45. Chen, Y.L., Hsu, C.C.: Self-regulated mobile game-based English learning in a virtual reality environment. Comput. Educ. **154**, 103910 (2020)

46. Dozio, N., et al.: A design methodology for affective Virtual Reality. Int. J. Hum.-Comput. Stud. **162**, 102791 (2022)

47. Dhimolea, T.K., Kaplan-Rakowski, R., Lin, L.: A systematic review of research on high-immersion virtual reality for language learning. TechTrends **66**(5), 810–824 (2022)

48. da Silva, R.L.: The process of moral decision-making in a game-based narrative scenario through the experience of future government workers. TechTrends **65**(4), 511–523 (2021). https://doi.org/10.1007/s11528-021-00591-y

49. Zheng, D., Newgarden, K., Young, M.F.: Multimodal analysis of language learning in World of Warcraft play: languaging as values-realising. ReCALL **24**(3), 339–360 (2012)

50. Deci, E.L., Ryan, R.M.: Intrinsic motivation and self-determination in human behaviour. Plenum Press, New York (1985)

51. Brophy, J.: Toward a model of the value aspects of motivation in education: developing appreciation for particular learning domains and activities. Educ. Psychol. **34**(2), 75–85 (1999)

52. Cummings, J.J., Tsay-Vogel, M., Cahill, T.J., Zhang, L.: Effects of immersive storytelling on affective, cognitive, and associative empathy: the mediating role of presence. New Media Soc. **24**(9), 2003–2026 (2022)

53. Hayes, A.T., Dhimolea, T.K., Meng, N., Tesh, G.: Levels of immersion for language learning from 2D to highly immersive interactive VR. In: Contextual Language Learning: Real Language Learning on the Continuum from Virtuality to Reality, pp. 71–89. Springer Singapore, Singapore (2021)

54. Gündüz, A.Y., Akkoyunlu, B.: Effectiveness of gamification in flipped learning. Sage Open **10**(4) (2020)

55. Ummihusna, A., Zairul, M.: Investigating immersive learning technology intervention in architecture education: a systematic literature review. J. Appl. Res. High. Educ. (2021)

56. Rey-Becerra, E., Barrero, L.H., Ellegast, R., Kluge, A.: The effectiveness of virtual safety training in work at heights: a literature review. Appl. Ergon. **94**, 103419 (2021)

57. Bourke, B.: Using gamification to engage higher-order thinking skills. In: I. Management Association (Ed.), Research Anthology on Developing Critical Thinking Skills in Students, pp. 632–652, IGI Global (2020)

58. Hsu, C.L., Chen, M.C.: Advocating recycling and encouraging environmentally friendly habits through gamification: an empirical investigation. Technol. Soc. **66**, 101621 (2021)

59. Chen, Y., Zhang, L., Yin, H.: A longitudinal study on students' foreign language anxiety and cognitive load in gamified classes of higher education. Sustainability **14**(17), 10905 (2022)

60. Dave, R.H.: Psychomotor levels. In: Armstrong, R.J. (ed.) Developing and writing behavioural objectives, pp. 20–21. Educational Innovators Press, Tucson (1970)

61. Ferris, T.L., Aziz, S.: A psychomotor skills extension to Bloom's taxonomy of education objectives for engineering education. Doctoral dissertation, National Cheng Kung University Tainan (2005)
62. Hughes, J.A., Prinz, W., Rodden, T., Schmidt, K., Robertson, T.: Cooperative work and lived cognition: a taxonomy of embodied actions. In: Proceedings of the Fifth European Conference on Computer Supported Cooperative Work, pp. 205–220. Springer Netherlands (1997)
63. Govender, T., Arnedo-Moreno, J.: An analysis of game design elements used in digital game-based language learning. Sustainability 13(12), 6679 (2021)
64. Argyriou, L., Economou, D., Bouki, V.: Design methodology for 360 immersive video applications: the case study of a cultural heritage virtual tour. Pers. Ubiquit. Comput. 24, 843–859 (2020)
65. Leenaraj, B., Arayaphan, W., Intawong, K., Puritat, K.: A gamified mobile application for first-year student orientation to promote library services. J. Librariansh. Inf. Sci. 55(1), 137–150 (2023)
66. Munafo, J., Diedrick, M., Stoffregen, T.A.: The virtual reality head-mounted display Oculus Rift induces motion sickness and is sexist in its effects. Exp. Brain Res. 235, 889–901 (2017)
67. Kilteni, K., Groten, R., & Slater, M.: The sense of embodiment in virtual reality. Presence: Teleoper. Virtual Environ. 21(4), 373–387, (2012)
68. Chang, C-C., Hwang, G-J.: An experiential learning-based virtual reality approach to fostering problem-resolving competence in professional training. Interact. Learn. Environ. (2021)
69. Ciubotaru, A.N., Devos, A., Bozorgtabar, B., Thiran, J.P., Gabrani, M.: Revisiting few-shot learning for facial expression recognition (2019)
70. Radianti, J., Majchrzak, T.A., Fromm, J., Wohlgenannt, I.: A systematic review of immersive virtual reality applications for higher education: design elements, lessons learned, and research agenda. Comput. Educ. 147, 103778 (2020)
71. Coban, M., Bolat, Y.I., Goksu, I.: The potential of immersive virtual reality to enhance learning: a meta-analysis. Educ. Res. Rev. 100452 (2022)
72. Dreger, K.C., Ticknor, B.: Situational XR: are there more than absolutes? J. Educ. Online 19(2) (2022)
73. Rizvic, S., Boskovic, D., Bruno, F., Petriaggi, B.D., Sljivo, S., Cozza, M.: Actors in VR storytelling. In: 2019 11th International Conference on Virtual Worlds and Games for Serious Applications (VS-Games), pp. 1–8. IEEE (2019)
74. Rizzetto, F., Rantas, S., Vezzulli, F., Cassin, S., Aseni, P., Vertemati, M.: New trends in surgical education and mentoring by immersive virtual reality: an innovative tool for patient's safety. In: Aseni, P., Grande, A.M., Leppäniemi, A., Chiara, O. (eds.) The High-risk Surgical Patient. Springer, Cham (2023). https://doi.org/10.1007/978-3-031-17273-1_58
75. Buyego, P., et al.: Feasibility of virtual reality-based training for optimising COVID-19 case handling in Uganda. BMC Med. Educ. 22(1), 274 (2022)
76. Dalim, C.S.C., Sunar, M.S., Dey, A., Billinghurst, M.: Using augmented reality with speech input for non-native children's language learning. Int. J. Hum Comput Stud. 134, 44–64 (2020)
77. Dreger, K.C., Ticknor, B.: Situational XR: are there more than absolutes? J. Educ. Online 19(2). https://doi.org/10.9743/JEO.2022.19.2.4, (2022)
78. Ahn, S.J.G., Nowak, K.L., Bailenson, J.N.: Unintended consequences of spatial presence on learning in virtual reality. Comput. Educ. 186, 104532 (2022)
79. Dubovi, I.: Cognitive and emotional engagement while learning with VR: The perspective of multimodal methodology. Comput. Educ. 183, 104495 (2022)

Intelligent Learning Systems for Simulator-Based Professional Training: A Systematic Literature Review

Martina Odéen[1], Charlott Sellberg[2], Gesa Praetorius[1,3(✉)], Oskar Lindwall[2], Linn Englund[1], and Anders Andersson[1]

[1] The Swedish National Road and Transport Research Institute (VTI), Regnbågsgatan 1, 417 55 Gothenburg, Sweden
`gesa.praetorius@vti.se`
[2] University of Gothenburg, Forskningsgången 6, 417 56 Gothenburg, Sweden
[3] University of South-Eastern Norway, Post Office Box 4, 43199 Borre, Norway

Abstract. This article presents the results of a systematic literature review on the use of Intelligent Learning Systems (ILS) in simulator-based professional education contexts aiming to provide guidelines on how to design a personalized, flexible, and adaptive learning experience to be used in remote simulator training for maritime cadets. The PRISMA protocol was employed to identify empirical studies published in international academic journals between 2018–2023. In the first step, 782 records were identified through searching through Scopus and Web of Science. After screening abstracts and full texts, a total of 10 studies remained. The article synthesizes results from three different domains where ILS have been used for simulator-based professional training: social science education, healthcare, and transportation. The results show a wide variety of applications and approaches to ILS design, such as to enhancing motivation and satisfaction in the students' learning process through adaptive feedback, or to provide real-time evaluations and feedback to trainees. The review shows the importance of considering end-user groups and how these may make use of the system in the design of an ILS. While systems can offer modules to both instructor and trainee, it is important to recognize that the two are different in terms of what is needed to facilitate an effective, efficient, and learner-centered training path.

Keywords: Intelligent Learning System (ILS) · Professional education and training · User-centered design

1 Introduction

In maritime education and training (MET), exercises in simulated environments play a central role for educating the future maritime workforce [1]. Recent advances in simulator technology offer opportunities for students to participate in remote training using cloud simulations [2]. However, a limitation of remote simulation training is that students typically do not receive personalized instruction, nor personalized feedback, and

T. E. Kim et al. (Eds.): MIS4TEL 2024, LNNS 1274, pp. 109–119, 2025.
https://doi.org/10.1007/978-3-031-84170-5_10

struggle to find the generic feedback from the system meaningful [3]. With the current advancements in training technologies, integrating Intelligent Learning Systems (ILS) with remote simulator training may provide new ways of addressing this problem, offering personalized feedback to students to help them learn in a meaningful and effective manner [4]. Within this study, the ILS is understood in accordance with [5] as a *"fully adaptive, personalized e-learning system without human teacher participation"*.

This study explores how ILSs have been integrated in simulator-based training in professional education by conducting a systematic literature review [6]. The review was designed to explore current developments of ILS that have been empirically tested in educational settings and focused on empirical studies published between 2018 and 2023 across a variety of settings where professional learning is in focus. With the aim to provide recommendations for the design of an ILS for MET, including guidance on how to develop a personalized, flexible, and adaptive learning experience in simulator-based maritime education, the following research questions guided the review: a) what kind of ILS designs have been developed for simulator-based professional training? b) what kind of learning theories guided these studies? c) how can lessons learned in other settings guide the development of an ILS for maritime simulator-based professional training?

2 Methodology

The systematic review method adopted in this study follows the Preferred Reporting Items for Systematic Reviews and Meta-Analyses [6]. The literature search was conducted in Scopus and Web of Science by the second author in October 2022, using a search string designed to include various notions of ILS and ILS platforms in the context of simulator-based training (Table 1). To focus the review on current advances in ILS research, the search for articles was limited to empirical studies published between 2018–2023 and filtered to highlight highly cited articles and popular topics in Web of Science. By only including articles in gold, silver, and bronze journals in Scopus, the study was limited to peer-reviewed articles in high-quality journals. Additionally, the search was focused on articles published in English to ensure relevance and transparency for an international audience.

Table 1. Search string

	intelligent OR smart OR adaptive OR personalized
AND	learn* OR train* OR assess* OR tutor*
AND	system* OR platform* OR environment
AND	simul*

The process of the screening, data collection and data items are illustrated in Fig. 1. As a result of the searches in Scopus and Web of Science, 782 articles were identified. The searches were saved in spreadsheets, containing information on bibliographic metadata and the abstract. Before screening, duplicates were removed (n = 1) as well as review articles found in the dataset (n = 4). Following this step, 777 abstracts were screened in a process where five of the six authors participated. Each abstract was read by at least two authors.

The inclusion criteria for articles were the following:

– the article presents empirical findings from the design and usage of an ILS,
– the ILS was designed and/or used within the setting of professional education and training,
– the focus of the article is not on the development of the algorithms within the system, but on how the system was actually used in practice.

At this stage, 606 records were excluded that did not meet the inclusion criteria; for example, by lacking a focus on a digitalized ILS, lacking a focus on professional education, or lacking an empirical foundation for the results presented. Data items for the remaining articles, in terms of the name of the study, year, authors, aim, method, intelligent learning systems, human performance measures, results, and discussion points, were summarized in a table. Next, 171 full texts were sought for retrieval. In this step, 4 articles could not be retrieved from the databases available through the university and were left out of the review. Hence, 167 full texts were assessed for eligibility by the first and second authors.

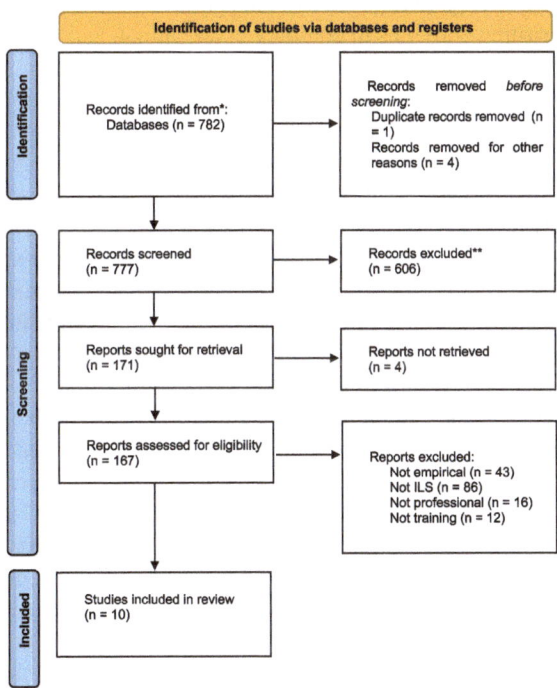

Fig. 1. Prisma protocol outlining the screening process.

In the last step, all articles not reporting empirical results, not being focused on professional education, i.e. studies within university settings or secondary schools were excluded, not containing an actual ILS, nor focusing on training were excluded. Only ten studies meeting our inclusion criteria remained.

3 Results

The review identified three domains in which ILS were used for professional training: Social science education (Sect. 3.1, n = 5), Healthcare (Sect. 3.2, n = 3), and Transportation (Sect. 3.3, n = 2).

3.1 Social Sciences Professional Education: Virtual Platforms, Automated Feedback, and Learning Analytics

Five studies report findings related to the use of ILS for professional education in the social sciences, including educational fields such as teacher training, psychology, and political science, where virtual learning environments have been combined with adaptive learning functions [8–11]. These studies mainly investigate ILS as a way of improving student motivation, satisfaction, and learning performance.

[9] developed an intelligent tutoring tool named EduZinc for the creation and assessment of student learning activities in complex open, online, and flexible learning environments. EduZing makes it possible for tutors to both create and assess course activities and materials. The tool can be used to create individualized learning products, such as activities and exams, automatically grade tests, create student, class, and competency-based reports, and deliver the reports to students, instructors, and tutors. An example of a more specialized platform is Enskill, a data-driven tool tailored for teaching English as a second language (L2 education) [10]. The Enskill prototype supports simulated exercises by evaluating content comprehension, pronunciation accuracy, and student responses, while also providing insights into exercise completion time, utterance duration, and conversational interactions. Both forms of feedback exhibited positive impacts on student satisfaction and learning outcomes; the former was found to enhance feedback quality, and the latter effectively evaluated student progress during dialogues [9]. In another study,[11] designed a serious game for training and assessing skills involved in classroom management, with a focus on non-technical skills such as negotiation, effective communication, and intercultural conflict management. The game is called "Attain Cultural Integration through Conflict Resolution Skill Development" (SG-ACCORD) and is a single-player virtual role-play, offering a series of scenarios reflecting the working life experiences of teachers. The interactions unfold between a player-controlled avatar (the teacher) and a computer-controlled counterpart (the student). SG-ACCORD's design aims to provide a user-friendly, adaptable educational resource that can stand alone, complete with automated feedback mechanisms and tools for post-simulation debriefing [11]. They [11] highlight how the systems log of interaction can facilitate teachers' retrospective self-reflections on the dynamics during the simulated activities. These reflections, it is argued, hold the potential to enhance teachers' conflict resolution skills.

Alternative methods involve the incorporation of adaptive features within existing frameworks. In [8] the authors illustrate this by integrating the programming software Scratch into a simulator system utilized for psychology education. The adaptive functions allow for managing the teaching software, monitoring psychological experimental processes as well as analysis of the students' work with psychological experiments. Another approach to automated feedback is to use learning analytics methods to assess skills in simulated environments [9, 12]. [12] explore the process of training negotiation skills in a collaborative simulation game used in a political science course. In this study, temporal social network analysis is used as a tool for understanding the students' learning trajectories over time and identifying the key moments when they need instructional support. [8] leverage learning analytics to scrutinize the correlations among 29 leadership skills, player strengths, and the flow experienced by participants during gameplay in a simulation named Fligby, designed to emulate leadership decision-making. Although Fligby does not offer direct automated feedback to players, it capitalizes on game-based psychometrics to furnish intricate insights into the interplay between players and the game.

3.2 Healthcare: Virtual Reality, Augmented Reality, and Conversational Agents

In healthcare education, ILSs have been introduced as a component in simulation-based surgical training. As noted by [13] and [14], there is a large and growing body of research studies showing the value of simulations for surgical training. Many simulation-based learning environments deliver feedback to the users, e.g., the amount of tissue damage, but according to [14], few simulators include any intelligent tutoring component. Ropelato et al. developed a custom-built system for the training of micromanipulation skills in Ophthalmic Surgery. The system consists of two parts: a) a virtual training environment enabled by an Augmented Reality (AR) headset (Microsoft Hololens) that visualizes the scenario and tracks the relevant micromanipulations; and b) a component that controls the administration of training tasks based on the performance of the users. The trainees interact with the system by using a microsurgical handle while observing a 3D holographic representation of the human eye and instrument position through the AR headset. The system continuously measures the performance of the trainees by logging the deviation from an optimal path. Five subtasks corresponding to different surgical procedures were developed, where each subtask could be offered with various levels of difficulty. The intelligent tutoring system was designed to continually match the difficulty level of the subtasks with the measured performance level, thereby offering the trainees tasks of appropriate challenge. By comparing the performance of a group of users that had the subtasks administered in sequential order with a group for whom the tasks were administrated by the intelligent tutoring system, the study shows that the ILS improves the overall results. With reference to Gallagher et al. (2012), the authors also raise the importance of objective, clear and complete metrics, such as the deviation from an optimal path in microsurgery.

As noted by [14] most surgical simulators are designed to train technical skills, such as micromanipulation techniques, and not non-technical skills like teamwork, communication skills and decision-making. Addressing this gap in research and development, [14] introduce what they call the first intelligent tutoring system for teaching surgical decision-making. Like the simulation investigated by [13], the system can be divided into two main comments: a) a virtual reality simulation consisting of a head-mounted display and haptic devices; and b) a conversational intelligent tutoring system called SDMentor (Surgical Decision-making Mentor). The system uses a standard intelligent learning system-framework including a domain model, student model, pedagogical module, and user interface. For the design of the domain model and pedagogical module, observational data of interaction between dental instructors and students were collected and analyzed with a focus on teaching strategies (e.g., asking questions, providing information, and giving feedback) and knowledge representation requirements (e.g., the effects of the performed actions). To evaluate the quality of the tutoring system, [14] compared it with tutorial feedback from experienced human tutors. The results showed that SDMentor was significantly better at providing tutoring interventions.

3.3   Transportation: Decision Support and Recommendation Systems

Two empirical studies using ILS approaches in the transportation domain were identified [15, 16]. [15] explore the use of mathematical modelling to develop an intelligent decision support and feedback system (IDSS) to enhance and improve helicopter simulator training, an "electronic instructor". The electronic instructor monitors the flight performance during the simulation and provides real time feedback enabling trainees to correct potential errors. The system aims to complement current simulation training approaches, which normally require feedback from instructors. As the IDSS is developed based on a mathematical model of the helicopters' flight dynamics and maneuverability, the system can both support and substitute the instructor during simulation exercises. The IDSS in the study is based on an expert system approach mimicking the pilot decision making at reference points to predict optimal control actions. Based on the prediction, deviations by trainees can be evaluated and appropriate instructions concerning flight paths, speed and altitude can be suggested. The authors argue that helicopter pilot training is mostly conducted in simulators, which makes it important to adapt instructional methods that can provide feedback that does not only correspond to what happens in the simulated world, but that corresponds to naturalistic flight environments and conditions.

[16] built a conceptual framework for driving assessment and recommendation systems to evaluate driving performance and guide driving behaviors with the aim to improve the network traffic conditions and drivers' perceived safety. Data mining techniques were used to extract, identify, characterize, and display driving behavior patterns. The pattern recognition system extracted information from raw data by identifying behavioral features such as lane changes, turns, harsh brakes etcetera. A Gaussian Mixture Model-Universal Background Model was then used to detect background factors from information of each driver in the dataset. At the same time an individual driving model for each driver is constructed to characterize an individual behavior profile. A comparison was then made between a driver's individual pattern and a standard "safe driver" pattern, which created a safety score. The scoring system was implemented in the drivers' cars and worked as a basis for assessing individual drivers, who were then recommended to mimic a nearby "safe" driver in a connected environment [16]. From the results it was shown that the proposed framework for behavior assessment and recommendation

system confirmed the capability to provide accurate detection, safety score calculation, and driving behavior characterization [16].

4 Summary of Findings

The findings of this review show that ILS as concept is currently still under development and can describe a wide range of different systems with different objectives and varying user groups. It is also noteworthy that only 10 out of a total number of 777 abstracts included in this study focused on empirical findings concerning the application of ILS in professional learning settings.

Synthesized from prior studies, prevailing learning models within the educational realm revolve around motivation, satisfaction, and active engagement [7–9]. The adaptive nature of ILS, grounded in the assimilation of student performance data, facilitates personalized guidance and real-time task difficulty adjustments to align with the learner's current knowledge level. This adaptability augments motivation and satisfaction for learners during task execution, bolstered by system-initiated notifications of task mastery [8] and tailored assistance during periods of challenge [12]. Active learning is palpable as students engage in reflective practices and respond to task-related inquiries during the process. The integration or fusion of ILS with virtual environments or software renders learning accessible and flexible, enabling students to complete assignments irrespective of their location or schedule. This feature is identified as a motivational attribute, enhancing the system's appeal [7, 8].

In the healthcare domain, active and tactile learning emerged as central models. These models were actualized using technologies such as virtual reality head-mounted displays [14] and augmented reality via tools like Hololens 2 and surgical instruments [13]. Within such setups, students engaged in hands-on practice of surgical techniques, and in instances like [14], were granted dedicated time for introspection after each subtask. The acquisition of input data for the Interactive Learning Systems (ILS) was multifaceted. For instance, [13] employed USB cameras and trainee-manipulated components on surgical instruments, while [14] captured cinematic data of the trainee's actions through virtual reality equipment. This real-time tracking and logging of trainees' surgical skills and decision-making furnished a wealth of information, empowering the ILS to provide instant adaptive feedback, whether during or post-training. This feedback encompassed instructions, guidance, task adaptation, and scoring adjustments, thereby enhancing the overall training experience.

Within the realm of transportation, ILS took on the roles of electronic instructors and recommendation systems. In this context, the ILS offered immediate adaptive feedback to steer trainees in their driving or flying endeavors [15, 16]. Given the inherent risks tied to driving and flying tasks, particularly the potential for accidents resulting from lapses in concentration, the application of active learning deviated from the patterns observed in prior domains. In the domains of education and healthcare, students had time to reflect and answer questions, while students in the transport domain instead followed instructions and safe error correction by the ILS based on their driving performance. In the study of helicopter flight training, data used to provide input to the ILS were flight parameters of a flight dynamic model representing the helicopter, as well as the trainee pilot's maneuvers and how well these corresponded to a desired move [15]. For car driving training, the input to the ILS was drawn from information extraction from the behavioral features of the driver, such as lane changes, harsh brakes, and turns [16].

5 Discussion

The history of ILS can be traced back to earlier work on computer-assisted instruction (e.g., [17]) and intelligent tutoring systems [18]. More recently, notions such as smart learning environments [19], intelligent learning management systems [20], adaptive learning systems (e.g. [21]), and personalized learning systems (e.g., [22]) have been introduced. Attempts to establish a common definition of ILS [19, p. 19] have been complicated by its complex history. Personalized and adaptive learning systems have been a topic for a long time [23] and incorporate knowledge and approaches across various research fields and disciplines, including computer science, psychology, and education [18]. It might therefore be hard to reach a consensus on central issues, such as which tools and features make a learning system intelligent and fit for specific purposes [20]. This lack of consensus is seen also in the results for this review, highlighting the diverse applications of Intelligent Learning Systems (ILS) within professional education across various domains. These applications range from tutoring systems for trainees to decision support systems aiding instructors in assessment. It is noteworthy that none of the studies presented findings from combining a tutoring and assessment functions, which may hint about the difficulty to design for multiple user groups at the same time as needs might be too diverse. As mentioned above, only 10 out of the 777 studies included empirical data about either the utilization of an ILS or evaluation. This shows that while there is a wide interest and range of publications, but many of these are on a conceptual or technical basis not actually addressing end-users and empirical findings from real-life settings.

Furthermore, as noted in the included studies, only one of the articles [9] utilized a pedagogical approach, i.e., the concept of *flow* to design the user interaction with the system. The other articles did either not explicitly mentioned any pedagogical approach or did not explicitly consider learners' perceptions and motivation in their system design approach. Additionally, many of the studies, such as [15] or [16] focused on offering feedback on specific metrics for isolated tasks or isolated situations, i.e. not considering learning advances over time. While this might be useful feedback on trainee performance during a single simulation and for the sake of assessment, it may not be what triggers learning and skill improvement, and may also explain the lack of reported research on achieving broader learning goals and outcomes. Questions persist regarding how system feedback contributes to increased performance at a higher level and how it aligns with learning goals and outcomes. The literature lacks clear answers on how performance pathways can be designed to allow learners to progress in multiple ways, especially in complex task environments like the maritime domain, where numerous potential actions may be possible in each situation. Clarifying these aspects remains a priority for future research in this field.

5.1 Limitations

The scope of the databases employed to find articles pertaining to Interactive Learning Systems (ILS) for simulator-based training was confined to Web of Science and Scopus. The adoption of additional databases might have yielded a more extensive array of articles, thereby potentially enriching the ensuing analysis. Moreover, by focusing solely on simulator-based training within professional education contexts, several potentially interesting studies were omitted. A case in point is the work of [24], which examined the efficacy of different feedback loops in an ILS. The integration of inner and outer loop feedback exhibited positive outcomes on student learning, prompting their inclusion in computer-based learning environments. Yet, [24] astutely points out that most preceding studies focused solely on one feedback type, thus failing to replicate the multifaceted feedback dynamic encountered in real-world teaching scenarios.

Another interesting point for discussion is the limitations of the results of the studies included in the review. Although most studies reported positive results of implementing an ILS in professional learning environments related to creative thinking, collaboration, and communication, as well as academic performance as a result of immediate feedback from the system [7, 8], several studies also reported limitations with the results in terms of internal and external validity. For example, in the studies of [8] and [14] the ILS were only tested for a short period of time, meaning that the improvement of the students' performance could not be determined with certainty whether the improvement was a consequence of using the tool or if other teaching related aspects were involved. Similarly, in the study of [13] the ILS was only tested on a specific surgical task, making it difficult to evaluate how well the system improves the student performance in other similar surgical tasks.

6   Conclusions and Implications

The published literature with empirical findings on the usage of ILS is up to now sparse. However, several important implications for the design of ILS in professional simulator-based education can be identified.

The end-users for the system need to be clearly identified early in the design process. Given that published literature only addresses either trainees or instructors as end-users indicates that the two groups may have different goals, motivations and information, as well as usability needs. These need to be thoroughly researched before the interaction and potential learning paths are identified and designed. Further, as simulator-based exercises within the maritime domain are conducted in groups, it is necessary to consider how to balance individual performance and group feedback. It is about both weighing what to assess and how to communicate this so that a learning experience including learning goals and learning achievements can be designed over time. While tailored solutions for groups are probably easier to achieve with regards to an incremental learning, it might be resource intensive and very complex to design learning paths for individual trainees embedded in groups during simulator exercises.

It is also important to consider the pace of feedback as well as how it is provided. ILSs reported about in this review almost exclusively focused on synchronous feedback instead of asynchronous feedback that triggers skill development over time. It indicates that there is a need for further research into learning paths and skill development within digital learning platforms. Furthermore, as learning is core to ILSs end-users, the system design should be built on solid pedagogical grounds to ensure that desired outcomes are achieved, not only in the short-term, but also in the long run. This may also include considerations on what type of data and information should be provided to ground the trainees' learning experience.

Acknowledgements. This research is funded by the European Union's Horizon Europe research and innovation programme under grant agreement No 101060107.

References

1. Wiig, A.C., Sellberg, C., Solberg, M.: Reviewing simulator-based training and assessment in maritime education: a topic modelling approach for tracing conceptual developments. WMU J. Marit. Aff. **22**(2), 143–164 (2023)
2. Kim, T.E., et al.: The continuum of simulator-based maritime training and education. WMU J. Marit. Aff. **20**(2), 135–150 (2021)
3. Gyldensten, W., Wiig, A.C., Sellberg, C.: Maritime students' use and perspectives of cloud-based desktop simulators: CSCL and implications for educational design. TransNav: Int. J. Marine Navig. Safety Sea Transport. **17** (2023)
4. Desmarais, M.C., d Baker, R. S.: A review of recent advances in learner and skill modeling in intelligent learning environments. User Model. User-Adap. Inter. **22**(1), 9–38 (2012)
5. Deliyska, B., Manoilov, P.: Ontologies in intelligent learning systems. In: Intelligent Learning Systems and Advancements in Computer-Aided Instruction: Emerging Studies, pp. 31–48. IGI Global (2012)
6. Moher, D., et al.: Preferred reporting items for systematic review and meta-analysis protocols (PRISMA-P) 2015 statement. Syst. Rev. **4**(1), 1–9 (2015)
7. Shen, C., Qi, A.: An adaptive learning mode of "public psychology" based on creative thinking with virtual simulation technology. Int. J. Emerg. Technol. Learn. **15**(23), 131–144 (2020)
8. Becerra-Alonso, D., Lopez-Cobo, I., Gómez-Rey, P., Fernández-Navarro, F., Barbera, E.: EduZinc: a tool for the creation and assessment of student learning activities in complex open, online, and flexible learning environments. Distance Educ. **41**(1), 86–105 (2020)
9. Buzady, Z., Wimmer, A., Csesznak, A., Szentesi, P.: Exploring flow-promoting management and leadership skills via serious gaming. Interact. Learn. Environ. 1–15 (2022)
10. Johnson, W.L.: Data-driven development and evaluation of Enskill English. Int. J. Artif. Intell. Educ. **29**(3), 425–457 (2019)
11. Dell'Aquila, E., Vallone, F., Zurlo, M.C., Marocco, D.: SG-ACCORD: designing virtual agents for soft skills training in the school context. Education Sciences **12**(3), 174 (2022)
12. Sun, Z., Theussen, A.: Assessing negotiation skill and its development in an online collaborative simulation game: a social network analysis study. British J. Educ. Technol. (2022)
13. Ropelato, S., Menozzi, M., Michel, D., Siegrist, M.: Augmented reality microsurgery: a tool for training micromanipulations in ophthalmic surgery using augmented reality. Simul. Healthcare **15**(2), 122–127 (2020)
14. Vannaprathip, N., Haddawy, P., Schultheis, H., Suebnukarn, S.: Intelligent tutoring for surgical decision making: a planning-based approach. Int. J. Artif. Intell. Educ. **32**(2), 350–381 (2022)
15. Guchenko, M., Shmakov, V., Yudina, A., Belska, V., Cejka, J., Bartuska, L.: An approach to developing mathematical software of on-board helicopter flight simulator decision support system. LOGI–Sci. J. Transp. Logist. **13**(1), 61–72 (2022)

16. Hong, Z., Chen, Y., Wu, Y.: A driver behavior assessment and recommendation system for connected vehicles to produce safer driving environments through a "follow the leader" approach. Accid. Anal. Prev. **139**, 105460 (2020)
17. Atkinson, R.C., Wilson, H.A.: Computer-assisted instruction. Science **162**(3849), 73–77 (1968)
18. Nwana, H.S.: Intelligent tutoring systems: an overview. Artif. Intell. Rev. **4**(4), 251–277 (1990)
19. Chen, N.S., Cheng, I., Chew, S.W.: Evolution is not enough: Revolutionizing current learning environments to smart learning environments. Int. J. Artif. Intell. Educ. **26**(2), 561–581 (2016)
20. Fardinpour, A., Pedram, M.M., Burkle, M.: Intelligent learning management systems: definition, features and measurement of intelligence. Int. J. Distan. Educ. Technol. **12**(4), 19–31 (2014)
21. Esichaikul, V., Lamnoi, S., Bechter, C.: Student modelling in adaptive e-learning systems. Knowl. Manage. E-Learn. Int. J. **3**(3), 342–355 (2011)
22. Shemshack, A., Spector, J.M.: A systematic literature review of personalized learning terms. Smart Learn. Environ. **7**(1), 1–20 (2020)
23. Tang, Y., Liang, J., Hare, R., Wang, F.Y.: A personalized learning system for parallel intelligent education. IEEE Trans. Comput. Soc. Syst. **7**(2), 352–361 (2020)
24. Tacoma, S., Drijvers, P., Jeuring, J.: Combined inner and outer loop feedback in an intelligent tutoring system for statistics in higher education. J. Comput. Assist. Learn. **37**(2), 319–332 (2021)

Instructional Approaches for Simulator-Based Maritime Education and Training

Franklin Nyairo[1]([⊠]), Per Haavardtun[2], Emilia Lindroos[1], Ziaul Haque Munim[2]([⊠]),
Jani Lampiola[1], Mirva Salokorpi[1], and Helene Krabbel[2]

[1] Faculty of Technology and Seafaring, Novia University of Applied Sciences, 20100 Turku,
Finland
franklin.nyairo@novia.fi
[2] Faculty of Technology, Natural, and Maritime Sciences, University of South-Eastern Norway,
3184 Horten, Norway

Abstract. This study explores the methods utilized by maritime instructors during simulator-based navigation exercises, particularly emphasizing their alignment with the International Convention on Standards of Training, Certification, and Watchkeeping for Seafarers (STCW). This study reveals that instructors are largely acquainted with traditional teaching methods such as direct instruction and scenario-based learning. Survey results indicate that a majority of instructors find their current approaches effective and compliant with STCW requirements. However, this study also uncovers areas of improvement, especially concerning learner engagement and the integration of modern technologies and best practices together with the adoption of innovative strategies such as adaptive learning and gamification. These findings hold significant implications for the design, implementation, and ongoing enhancement of maritime training programs, aiming to contribute towards the more effective educational practices in the maritime sector to foster the competence and safety of maritime professionals.

Keywords: Simulator-based training · maritime education and training · competency-based learning · pedagogical approaches · experiential learning

1 Introduction

Simulator-based learning has revolutionized maritime education and training (MET) by providing a realistic and risk-controlled environment for learners [1]. Simulators in maritime education and training align with the International Convention on Standards of Training, Certification, and Watchkeeping for Seafarers (STCW), emphasizing practical training and competence development [2]. Teaching models in MET are anchored on behaviorist, cognitive, social-interaction, personalized learning, and constructive theories. One primary model implemented in MET is the experiential learning model, an approach rooted in constructivism [3].

MET institutions need to comply with the International Maritime Organization (IMO) regarding what competence trainees must learn during the program [4]. Usually, a MET institution offering a navigation bachelor program utilizes simulators for

T. E. Kim et al. (Eds.): MIS4TEL 2024, LNNS 1274, pp. 120–131, 2025.
https://doi.org/10.1007/978-3-031-84170-5_11

RADAR, Automatic Radar Plotting Aids (ARPA), Electronic Chart Display and Information System (ECDIS), Global Maritime Distress Safety System (GMDSS), and for ship handling or navigational training. There are four different simulator modalities – full mission, desktop, cloud, and virtual reality (VR) simulators [2, 5]. While desktop simulators can run on computers, full mission simulators replicate a ship's bridge in high fidelity with the support of VR [2].

This study investigates the familiarity of MET instructors with various teaching and learning methods, and the instructors' perceptions of the use and impact of these instructional approaches. Data was collected from MET instructors in three Nordic institutes. The findings suggest that instructors are highly familiar with direct instruction or lectures, briefing-debriefing, and providing detailed feedback as core instructional strategies.

The rest of the study is structured as follows: Sect. 2 presents the theoretical background for the research. Section 3 explains the methodology used. Section 4 presents the findings which are elaborated on in Sect. 5. Section 6 summarizes with the conclusion and future research.

2   Literature Review

In the teaching and learning process, an understanding of the theoretical frameworks underpinning instructional approaches is relevant to ensure teaching and learning objectives are addressed. Central to this endeavor is the exploration of learning theories—namely behaviorism [6], cognitivism [7], constructivism [8], and connectivism [9]—which seek to elucidate the mechanisms through which learning takes place, thereby forming the bedrock for developing strategies.

The subsequent literature review explores the extensive array of instructional strategies and methods integral to the MET instructional framework. The Saskatchewan Education project delineates instructional strategies into five categories: Direct Instruction, Indirect Instruction, Experiential Instruction, Independent Study, and Interactive Instruction [10].

2.1   Instructional Strategies

Instructional strategies in MET are diverse, ranging from traditional to cross-disciplinary and activity-based approaches. One traditional prevalent strategy is direct instruction, which focuses on the provision of information and step-by-step skill development. However, more contemporary methods like problem-based learning are becoming integral.

In problem-based learning, learners engage with real-world scenarios such as navigating a ship in low visibility, thereby fostering their critical thinking skills and theoretical knowledge to practical contexts [12]. One challenge is to make sure that the training results in the learning experience in such a way that the students will can make use of it in the future, also defined as positive transfer of training [13, 14]. The idea is to design the scenarios as realistic as possible, not only in the physical sense but more importantly in the roles the students take on.

2.2 Instructional Methods

Various instructional methods are employed to deliver maritime education. The traditional lecture method is still significant in imparting theoretical knowledge, such as maritime law or marine engineering principles [15]. Alongside, the demonstration method allows learners to observe specific skills under expert supervision, like the usage of radar or ECDIS in simulators [16]. Other methods like drill and practice, the project method, the case study method, and the scenario-based learning offer unique learning opportunities. For instance, the drill and practice method aids in mastering technical skills like knot-tying or radio communication equipment usage, while project and case study methods foster problem-solving and teamwork skills through in-depth incident analysis [1]. But these basic skills will not be of any help for the students in their future work if they are not put together and applied in a real-world setting. This setting can be provided by a simulator but requires the instructors to maintain an updated knowledge base. This is the competency-based training idea, which facilitates the connection between the teaching and the workplace environment [17].

An array of teaching techniques can be employed in MET to augment these instructional methods. Techniques such as interactive questioning and group discussions stimulate critical thinking [18]. Role-playing exercises can offer learners a deeper understanding of different roles within a ship's crew, fostering communication and cooperation skills [3].

To summarize, an effective blend of instructional strategies and methods is critical for ensuring a comprehensive learning environment in MET [19]. However, the effectiveness of these pedagogical elements can vary based on learning objectives, learner characteristics, and contextual factors, underscoring the need for continuous pedagogical research and innovation in response to the dynamic nature of the maritime industry [3]. Table 1 presents the list of instructional strategies and their underlying methods explored in this study.

3   Method

3.1 Data Collection

To address the research objective, this study adopted a survey-based design. Data was collected using a survey which was administered online from May to June 2023 among maritime instructors involved in simulator-based navigation training at three Nordic MET institutions. In total, 14 complete responses were received. Table 2 presents an overview of the respondents' background.

Table 1. Common instructional strategies and methods

Strategy	Method	Description
Teacher-Directed	Direct Instruction/ Lectures	Teacher/Instructor is the primary source of information, delivered through lectures
Learner-Centered	Learner-Centered	Teacher/Instructor acts as a facilitator, students work on scenarios and reflective discussion
	Independent Study and Practice	Students take responsibility for their own learning through self-study or practice
Interactive/ Experiential Learning	Interactive and Experiential Learning (Simulations, Role Plays)	Combines interactive instruction (simulations) with experiential learning (role plays) for a more immersive experience [3, 20]
	Scenario-Based Learning	Students are presented with specific scenarios or problems to solve [21]
	Gamification	Uses game elements in non-game contexts to motivate and engage students
	Collaborative Learning	Encourages students to work together in groups to solve problems or complete tasks [18, 22]
Personalized learning	Adaptive Learning	Uses technology to adjust the pace and level of instruction to individual student needs [23]
Supportive Strategies/Indirect Instruction	Briefing and Debriefing	Briefing sets the stage for learning, debriefing involves reflecting on the activity after completion
	Providing Feedback	This is an essential part of learning that involves giving information about performance to aid improvement

Table 2. Background of respondents

Variable	Category	Frequency	Percentage
Institution	Institution A	6	43%
	Institution B	4	29%
	Institution C	4	29%
	Total	14	
Age Group			
	20–29	0	0%
	30–39	5	36%
	40–49	3	21%
	50–59	4	29%
	over 60	2	14%
	Total	14	100%
Teaching Experience (years)			
	less than 5	2	14%
	6–10	2	14%
	11–15	4	29%
	16–20	3	21%
	over 20	3	21%
	Total	14	100%

3.2 Survey Design

The survey included questions regarding the familiarity, use, and impact of several instructional strategies. A list of 10 instructional approaches were assessed, which were adopted from [10]. The survey questions can be grouped into three categories:

- **Familiarity with instructional approaches**: To assess familiarity of various instructional approaches that instructors employ in simulator-based navigation exercises.
- **Usefulness of instructional approaches**: To assess instructors' perception of the usefulness of the instructional approaches they use.
- **Impact of instructional approaches**: To explore instructors' perception of how different instructional approaches impact the performance of the learners in simulator-based navigation exercises.

The survey was designed to capture the data on a 5-point Likert-scale, where 1 referred to not familiar at all or strongly disagree and 5 to extremely familiar or strongly agree. Data collected through the survey was analysed using descriptive statistics and various plots based on the frequency of the responses. SPSS and Microsoft Excel were used for the analysis.

4   Results

4.1   Familiarity with Instructional Approaches

The degree of familiarity with various instructional approaches can indicate instructors' over-dependency and under-utilization of approaches in MET. The mean values and standard deviation of the 14 respondents on the familiarity with instructional approaches are reported in Fig. 1.

The results indicate a high level of familiarity among maritime instructors with Briefing and Debriefing (mean = 4.21, sd = 0.70) and Direct Instruction/Lectures (mean = 4.21, sd = 0.58). However, there is a noticeable lack of familiarity with more innovative strategies such as Adaptive Learning (mean = 2.79, sd = 1.05) and Gamification (mean = 2.29, sd = 0.83). This suggests an opportunity for exploring the potential of these instructional approaches in MET.

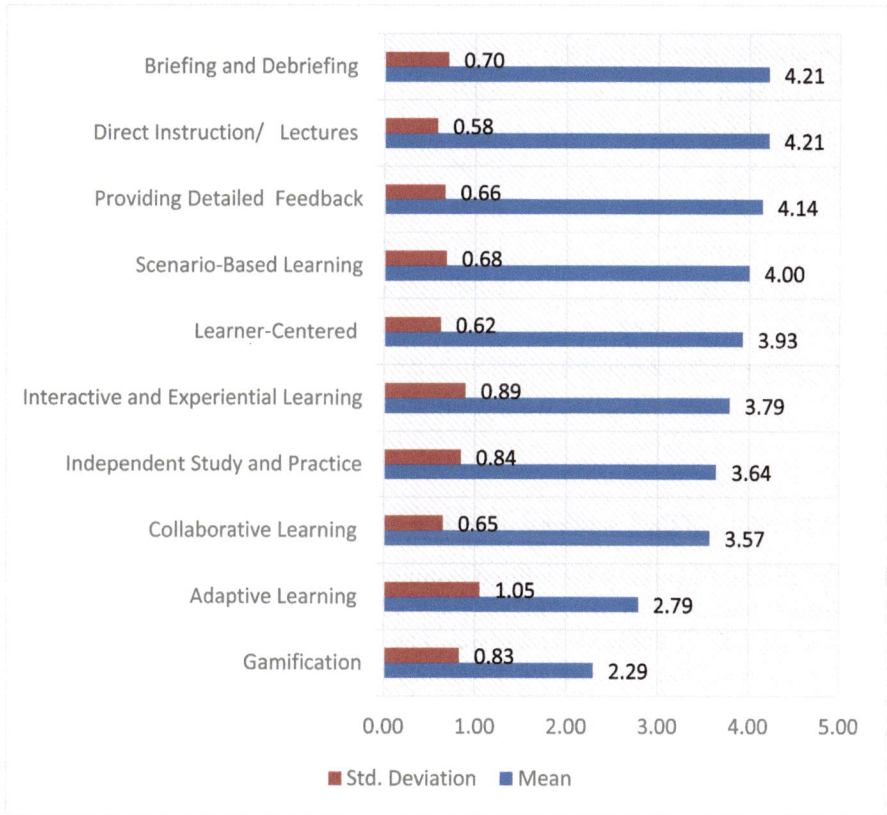

Fig. 1. Instructors' familiarity with various instructional approaches (N = 14)

Further, the Independent-Samples Kruskal-Wallis Test was used to explore differences in the degree of familiarity of the 10 instructional approaches across the three MET institutions, the results can be seen in Table 4. Kruskal–Wallis is used since differences across more than two conditions (herein, institutions) are explored, and data is assumed nonparametric due to sample size. At 5% statistical significance, differences were found across two instructional approaches—*Independent Study and Practice* and *Briefing and Debriefing*. Further, at 10% statistical significance, differences were found in *Adaptive Learning*. Boxplot diagrams are explored (available upon request) demonstrating the differences across institutions. According to the boxplots, *Independent Study and Practice* is more utilized in institution A and C, *Briefing and Debriefing* in institution A and C, and *Adaptive Learning* in institution C.

Table 3. Mean comparison test result of degree of familiarity across institutions

Instructional Approach	P-value	Decision at 5% Statistical significance*
Direct Instruction/Lectures	0,569	Retain the null hypothesis
Learner-Centered	0,206	Retain the null hypothesis
Interactive and Experiential Learning	0,403	Retain the null hypothesis
Independent Study and Practice	*0,031*	*Reject the null hypothesis*
Scenario-Based Learning	0,637	Retain the null hypothesis
Adaptive Learning	*0,097*	*Retain the null hypothesis*
Gamification	0,684	Retain the null hypothesis
Collaborative Learning	0,881	Retain the null hypothesis
Briefing and Debriefing	*0,040*	*Reject the null hypothesis*
Providing Detailed Feedback	0,793	Retain the null hypothesis

* *Null hypothesis: The distribution of the instructional approach is the same across the three institutions.*

4.2 Perceptions of the Usefulness of Instructional Approaches

The perceived usefulness of the utilized instructional approaches by the instructors is examined in four dimensions: appropriateness, effectiveness, improving learners' abilities, and addressing learners' weaknesses. Figure 2 presents the mean and standard deviation values of the four dimensions. Among the four dimensions, improving learners' abilities has the highest score (mean = 3.93, sd = 0.62) followed by effectiveness, appropriateness, and addressing learner weaknesses.

To explore variations in the perceived usefulness across institutions, a radar diagram with the mean values of the four dimensions across the institutions is explored (available upon request). It is observed that the institutions vary to a great extent in two of the usefulness dimensions—*address the learners' areas of weakness* and *effectiveness* of the instructional approach. While institution C seems to have the highest mean value on *address the learners' areas of weakness*, it has the lowest on *effectiveness*.

4.3 Perception of the Impact of Instructional Approaches

Finally, the perceived impacts of the utilized instructional approaches are examined in four dimensions: easiness of implementation, flexibility, increased learner engagement, and improvement of overall quality. According to the mean values presented in Fig. 3, MET instructors perceive their use of instructional approaches easy to implement (mean = 4.07, sd = 0.48) followed by improvement of overall quality, flexibility, and increased learner engagement.

To explore differences on the perceived impact of instructional approaches, mean scores on the four dimensions are explored in a radar diagram (available upon request). While the institutions are rather aligned, some differences are observed in the flexibility dimension, where institutions B and C have higher scores than the overall average, and institution A slightly lower.

Fig. 2. Perception of usefulness (N = 14)

5 Discussions

The findings suggest that the instructors are familiar with the traditional instructional approaches, that is, the teaching-and-learning paradigm [24]. Simulator training focuses on briefing and debriefing and one-to-one instructions during the exercises [25]. The lowest score is obtained by gamification and adaptive learning, which are relatively new instructional approaches and not something broadly adopted by the maritime teaching community yet.

The focus on increasing the learners' ability is seen as the most useful. This is in line with the simulator syllabus, which requires giving the students the right abilities

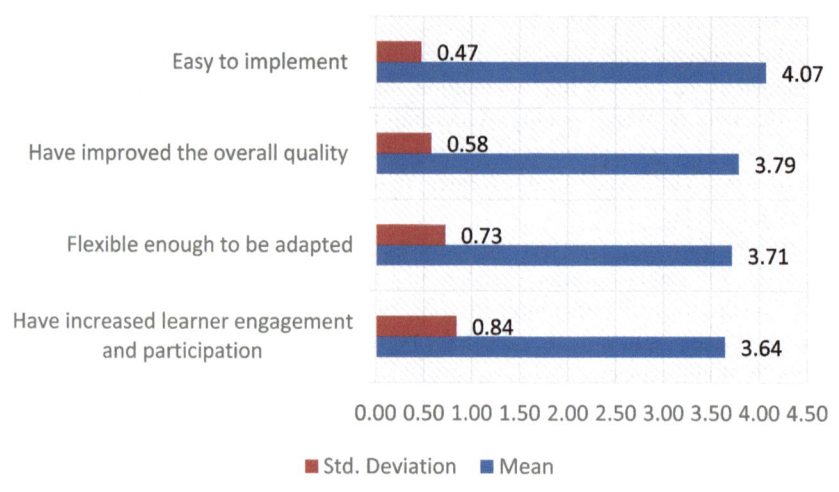

Fig. 3. Perceived impact of instructional approaches (N = 14)

to conduct safe navigation. The lowest score is on addressing the learners' weaknesses. This can possibly be explained by having larger groups in simulator training rather than one on one teaching, thus making it more difficult to address weaknesses of single students. Hence, the focus is on the effective and appropriate instructions in order to fulfill the learning objectives, as seen in [26]. A comparative study on the usefulness of instructional approaches considering the number of students, instructors and available simulators should be researched in future studies.

The factor 'ease of implementation' scored the highest, which can be attributed to the time constraints of instructors. Given their limited spare time, experimenting with new techniques is often time-consuming. Therefore, solutions that offer a noticeable increase in effectiveness are typically adopted [27].

6   Conclusions

This study explored the instructional approaches adopted by maritime instructors within simulator-based navigation exercises. The primary objective was to discern the extent to which traditional and innovative instructional strategies are employed, and how these align with international maritime training standards. Survey data was collected from instructors across three Nordic maritime education and training institutions.

The findings explicitly affirm that maritime instructors exhibit a strong familiarity with traditional instructional methodologies and express a belief in their effectiveness and congruence with STCW stipulations. The findings also unveil areas needing further research, notably the incorporation of innovative instructional strategies and a more targeted approach to enhance learners' engagement within the learning environment. Despite the prevalent reliance on well-established teaching methodologies, there exists room for the inclusion of adaptive learning and gamification techniques, which are presently underutilized within the surveyed institutions.

A potential limitation of this study is the relatively modest sample size, even though it is representative of the instructor population across the participating institutions. The statistical analysis employed non-parametric tests to mitigate the impact of the sample size. Moreover, there is potential for several future research directions. It would be relevant to extend this inquiry to a broader array of MET institutions, possibly encompassing varied geographical and institutional contexts, to garner a more holistic understanding of instructional strategies in simulator-based maritime training on a global scale. Furthermore, exploratory studies focusing on the effective integration of innovative instructional techniques such as adaptive learning and gamification within the maritime education framework could potentially unveil novel pathways for enhancing the quality and effectiveness of maritime training.

Informed Consent. All respondents involved in this study provided informed consent prior to participation, and provided written and/or verbal agreement. They were adequately informed about the nature, purpose, and potential impacts of the research, and were also apprised of their rights to confidentiality and withdrawal from the study at any stage without any repercussions.

Acknowledgments. The research is part of the Integrating Adaptive Learning in Maritime Simulator-Based Education and Training with Intelligent Learning System (I-MASTER) project supported by the European Union's Horizon Europe research and innovation programme under grant agreement No. 101060107.

Disclosure of Interests.   The authors declare that there is no conflict of interest regarding the publication of this paper. The research conducted and presented herein was carried out independently, and there are no financial or personal relationships with other individuals or organizations that could have inappropriately influenced the work. The authors have adhered to all institutional guidelines and ethical standards in the conduct and dissemination of research.

References

1. Wiig, A.C., Sellberg, C., Solberg, M.: Reviewing simulator-based training and assessment in maritime education: a topic modelling approach for tracing conceptual developments. WMU J. Marit. Affairs **22**(2), 143–164 (2023). https://doi.org/10.1007/s13437-023-00307-4
2. Kim, T., et al.: The continuum of simulator-based maritime training and education. WMU J. Marit Affairs **20**(2), 135–150 (2021). https://doi.org/10.1007/s13437-021-00242-2
3. Hanzu-Pazara, R., Barsan, E., Arsenie, P., Chiotoroiu, L., Raicu, G.: Reducing of maritime accidents caused by human factors using simulators in training process. J. Maritime Res. **5**(1), 3–18 (2008)
4. International Maritime Organization, 'International Convention on Standards of Training, Certification and Watchkeep-ing for Seafarers' (2018) Please check and confirm if the authors given and family names have been correctly identified (2023). https://www.imo.org/en/OurWork/HumanElement/Pages/STCW-Convention.aspx
5. Tusher, H.M., Munim, Z.H., Nazir, S.: An evaluation of maritime simulators from technical, instructional, and organizational perspectives: a hybrid multi-criteria decision-making approach. WMU J. Marit. Affairs (2023). https://doi.org/10.1007/s13437-023-00318-1

6. Skinner,B.F.; Science and human behavior (No. 92904). Simon and Schuster New York, NY (1953)
7. Piaget, J.: The origins of intelligence in children. W W Norton & Co, New York (1952). https://doi.org/10.1037/11494-000
8. Vygotskij, L.S., Cole, M.: Mind in society: the development of higher psychological processes, Nachdr. Harvard Univ. Press, Cambridge, Mass (1981)
9. Siemens, G.: Connectivism: a learning theory for the digital age. Int. J. Instruct. Technol. Distan. Learn. 2(1), 3–10 (2005)
10. S. S. Education, Instructional Approaches: a Framework for Professional Practice. Saskatchewan Education (1991). https://books.google.de/books?id=kAMyNAEACAAJ
11. Joyce, B.R., Weil, M., Calhoun, E.: Models of teaching, Ninth edition. Upper Saddle River: Pearson (2017)
12. Fedila, M.: Appropriateness of problem based learning in maritime education and training (2007). https://api.semanticscholar.org/CorpusID:55581860
13. Newstrom, J.W.: role-taker/time differential integration of transfer strategies (1984)
14. Grossman, R., Salas, E.: The transfer of training: what really matters. Int. J. Train. Develop. 15(2), 103–120 (2011). https://doi.org/10.1111/j.1468-2419.2011.00373.x
15. Abeysiriwardhane, A.: Learning and learning-to-learn by doing: an experiential learning approach for integrating human factors into maritime design education. Marit. Technol. Res. 3(1) (2020). https://doi.org/10.33175/mtr.2021.241912
16. Lützhöft, M., Nyce, J.M.: Integration work on the ship's bridge. J. Marit. Res. 5(2), 59–74 (2008)
17. Harris, R., (ed.): Competency-based education and training: Between a rock and a whirlpool, Repr. South Yarra: Macmillan Publ. Australia (2001)
18. Jones, J.M.: Discussion group effectiveness is related to critical thinking through interest and engagement. Psychol. Learn. Teach. 13(1), 12–24 (2014). https://doi.org/10.2304/plat.2014.13.1.12
19. Zhu, M., Berri, S., Zhang, K.: Effective instructional strategies and technology use in blended learning: a case study. Educ. Inf. Technol. 26(5), 6143–6161 (2021). https://doi.org/10.1007/s10639-021-10544-w
20. Chi, M.T.H.: Active-constructive-interactive: a conceptual framework for differentiating learning activities. Top. Cogn. Sci. 1(1), 73–105 (2009). https://doi.org/10.1111/j.1756-8765.2008.01005.x
21. Salas, E., Wilson, K.A., Burke, C.S., Priest, H.A.: Using simulation-based training to improve patient safety: what does it take? Joint Comm. J. Qual. Patient Safety 31(7), 363–371 (2005). https://doi.org/10.1016/S1553-7250(05)31049-X
22. Dillenbourg, P., Järvelä, S., Fischer, F.: The evolution of research on computer-supported collaborative learning. In: Balacheff, N., Ludvigsen, S., De Jong, T., Lazonder, A., Barnes, S. (eds.) Technology-Enhanced Learning, pp. 3–19. Springer Netherlands, Dordrecht (2009). https://doi.org/10.1007/978-1-4020-9827-7_1
23. Vandewaetere, M., Desmet, P., Clarebout, G.: The contribution of learner characteristics in the development of computer-based adaptive learning environments. Comput. Hum. Behav. 27(1), 118–130 (2011). https://doi.org/10.1016/j.chb.2010.07.038
24. Ruben, B.D.: Simulations, games, and experience-based learning: the quest for a new paradigm for teaching and learning. Simul. Gaming 30(4), 498–505 (1999). https://doi.org/10.1177/104687819903000409
25. Sellberg, C.: From briefing, through scenario, to debriefing: the maritime instructor's work during simulator-based training. Cogn Tech Work 20(1), 49–62 (2018). https://doi.org/10.1007/s10111-017-0446-y

26. Karahalil, M., Lützhöft, M., Scanlan, J.: Formative assessment in maritime simulator-based higher education. WMU J Marit Affairs **22**(2), 181–207 (2023). https://doi.org/10.1007/s13437-023-00313-6

27. Jamil, M.G., Bhuiyan, Z.: Deep learning elements in maritime simulation programmes: a pedagogical exploration of learner experiences. Int. J. Educ. Technol. High. Educ. **18**(1), 18 (2021). https://doi.org/10.1186/s41239-021-00255-0

Author Index

© The Editor(s) (if applicable) and The Author(s) 2025
T. E. Kim et al. (Eds.): MIS4TEL 2024, LNNS 1274, p. 133, 2025.
https://doi.org/10.1007/978-3-031-84170-5